Evolution's Way

Toward Exponentially Higher States of Complexity, Consciousness, and Unity

———

Todd F. Eklof

Evolution's Way

Toward Exponentially Higher States of Complexity, Consciousness, and Unity

---

Todd F. Eklof

Oakleaf Press
P.O. Box 19542
Spokane, WA, 99224

Copyright © 2020 by Todd F. Eklof
All Rights Reserved
ISBN: 9798626653793

To my friend Dennis Held, a poet, writer, and my editor, who was diagnosed with stage-4 melanoma just prior to working on this book and beat the odds while doing so thanks to exponential advances in medicine and his enduring spirit.

Contents

INTRODUCTION	i
A LETTER TO THE FAR FUTURE	1
AIMING FOR TOMORROW	12
AUTOBIOGRAPHY OF THE UNIVERSE	27
COMFORT MACHINES	38
DIFFERENTIATION, MY DEAR WATSON	52
THE ONCE AND FUTURE RESURRECTION	65
WAKING BRAHMAN	79
DOOMING DOOMSDAY	90
THE EVOLUTION OF ONENESS	102
THE SINGULARITY AND ME	110
CHANNELING YOUR INNER THERAPIST	125
MY MOONSHOT HEROES	139
FROM RICHES TO RAGS	153
THE GOOD NEWS	165
FACING VIRTUAL REALITY	180
NOTHING BUT TAXES	195
THE HOUSE OF HUMANITY	210
LEAVING EARTH	225
THE TOOLMAKER'S PARADOX	237
AFTERWORD	248
ACKNOWLEDGMENTS	254
BIBLIOGRAPHY	257
INDEX	263

INTRODUCTION

This book is not a scientific treatise and I am not a scientist. I'm a minister and it's based on a collection of sermons I've given over the course of fifteen years, beginning in 2004. They're founded on my understanding of evolution as *the exponential flow of everything—life, ideas, society, technology, humanity, and the entire Cosmos—toward increasingly higher states of complexity, consciousness, and unity: in that order*. Each touch upon a recurring theme in my pulpit: that the Universe itself is awakening, and *Homo sapiens* represents a significant bifurcation in this evolutionary process. This is so, largely, because human

INTRODUCTION

beings, to a much greater degree than any other animal on Earth, have developed the ability to exchange information through the sharing of ideas, freeing it from its genetic confines. This book, like any book, is a good example. Information can now be exchanged between host organisms in the form of memes as well as genes.

Human technology is also evolving exponentially to facilitate this obvious trajectory. I need only mention computers and the Internet to prove the point. Advances now underway in robotics, extended reality, and artificial intelligence are also on the verge of expanding human intelligence and intelligence itself in the coming years—much sooner, I think, than most imagine. As we increasingly converge with such technology, our species, like 99 percent of all species to have come before us, will go extinct, likely by evolving into something new. I predict the human part of us will remain within whatever we become, just as our single-celled ancestors remain part of us. Human *technology*, from the Greek word meaning "art" or "craft," is an extension of our humanity, so there's already something human in all of it.

I realize such talk frightens some. I've seen all the *Terminator* films and am well aware of the modern "Rise of the Machines" mythos. The original sermons this book is based on weren't always fully appreciated by those who heard them, either. If we can be sure of anything in this world, most would agree that machines are not alive, let alone human—that they are the complete opposite of what it means to be human. I disagree and take on this specific paradigm in the final chapter, although my optimistic bias for our mechanical descendants is apparent throughout. You may or may not agree with the reasons presented for this positive outlook, but I do my best to explain why I think it's warranted.

Indeed, as I've taken on the topic over time, my sermons themselves have evolved in parallel with exponential advances in technology. I divide the chapters into two parts, representing two phases of my thinking: Pre-SU and Post-SU. SU stands for Singularity University, a Silicon Valley-based Benefit corporation cofounded by futurist Ray Kurzweil and entrepreneur Peter Diamandis. SU's mission is to educate global leaders about the opportunities of harnessing the power of exponentially improving technologies to help solve some of the world's biggest challenges.

I first learned of Singularity University while reading of it in Diamandis's inspiring 2012 book, *Abundance: The Future is Better than You Think*, coauthored with his writing partner, Steven Kotler. "What's needed is a place where people can go to hear of the biggest and boldest ideas, those exponential possibilities that echo Archimedes: 'Give me a

INTRODUCTION

lever long enough, and a place to stand, and I will move the world.'"[1] Having already become aware and excited about the crucial role human technology is playing in the evolution of consciousness, my yearning began. I saved my pennies and was finally able to complete SU's winter Executive Program in 2018, one of the most valuable and transformative experiences of my life.

Prior to this, my sermons approached the subject more philosophically, an important foundation for appreciating the Abundance mindset, the promise of exponentially evolving technologies. For this reason, it is my hope this book will be of special value to my fellow SU graduates and members of my Abundance community. The promise of Abundance was born in the stars 13.7 billion years ago and has been exponentially moving in that direction ever since. If, as many religions suggest, having inner peace means being at one with the ways of the Universe, at one with *The Way*, then it may be that we should all do our best to move toward a world of abundance. Maybe this is what Jesus understood when he asked his followers to help create "Heaven on Earth," and what he meant when he said, "I have come that they may have *life*, and that they may have it more *abundantly*." It's certainly why, as a minister, unorthodox as I am, this mindset moves me deeply and has become foundational to my spiritual life.

This explains my title, *Evolution's Way*. Evolution is a spiritual force as much as it is a scientific principle: the two are not mutually exclusive. In the martial arts world, the Japanese word *Do*, or *Tao* in Chinese, means something like The Path, The Art, or The Way. A *Dojo*, where martial arts are practiced, means, "Place of the Way." *Aikido* means, "The Harmonious Way," *Judo*, "The Gentle Way," *Iaido*, "The Prepared Way," and *Budo*, "The Warrior Way." Tao, as in *Taoism*, simply means, "The Way." In this sense, Evolution may be synonymous with *Tao*, The Way, for it is the path of *all* things, the path of the Universe itself.

Although many in the religious world have long recognized this larger picture, that we are an intricate part of an awakening Universe, few have appreciated—and some resist—the idea that our machines are a vital and beautiful part of this process, of Evolution's Way. Many view machines as the antithesis of being human, not an extension of it, and our technologies as necessary evils, but not art. For these, I hope my post-SU chapters, which place a greater emphasis upon the advancement of specific technologies, may promote a shift around such thinking.

All technologies are *human* technologies and can be used to ease our burdens or do terrible things. The pitchfork made for baling hay can as easily be used as the favored

INTRODUCTION

weapon of a mob, just as today's social media can. A knife meant for cutting meat and slicing cheese is deadly in the hands of a murderer. Nuclear power can be harnessed to energize cities or demolish them. The problem is not technology but how we use it, to advance human welfare and fulfillment, or for destruction. The Abundance mindset recognizes this distinction and is motivated by the humanistic ethic, including proper care for our planet and fellow creatures. Promoting Abundance isn't just about promoting technology, but about its proper purpose and use. Abundance is about making the right choices now in light of emerging technological certainties.

Finally, I should probably say a little about my peculiar religion, to help explain the unusual nature of my sermons, which, for many, won't sound like sermons at all. Unitarian Universalism is rooted in theological ideas dating back to 1st century Christianity, based upon the belief that Jesus was an admirable human being, but not a god. Over time this humanistic Christology, especially as it has manifested in the United States, influenced greatly by Enlightenment philosophy, has become nonsectarian, nontheistic, noncreedal, and, as our Associational bylaws explain, embraces "Humanist teachings which counsel us to heed the guidance of *reason* and the results of science, and warn us against idolatries of the *mind and spirit*."

As a noncreedal religion, not unlike Buddhism, or Taoism, or Hinduism, Unitarian Universalists have no dogma or specific theology we're required to believe. Some believe in God, others are atheists, and everything in between. As individuals, we are free to draw inspiration from all the world's religions and philosophies, or from none. Our congregations congregate not because we hold the same beliefs, but because we share similar values about how we ought to treat each other, values I might summarize as dignity, equality, openness, freedom, democracy, justice, and unity. But even these are common human values shared by many.

So you won't find a lot of traditional religious jargon in the following chapters. They aren't about promoting my faith or converting my reader since an important part of my religion is valuing your right to discover and pursue your own path, your own Way. Rather, they are about my own intellectual pilgrimage over the course of the past several years, during which I have become more optimistic about our species and its future. Bertrand Russell once said, "A complete philosopher will have a conception of the ends to which life should be devoted, and will be in this sense religious."[2] Religion for any of us is defined by what we

INTRODUCTION

devote our lives to, and we aren't complete if we're not devoted to something. Devoting everything we do, including our technologies, to advance life and human wellbeing seems a worthy endeavor, whether we call such devotion our religion, our massively transformative purpose, or simply that which gives our lives meaning. I wrote the following sermons, messages, essays, chapters, whatever you wish to call them, because they mean something to me. I hope they might also mean something to you.

---

[1] Diamandis, Peter H., and, Kotler, Steven, *Abundance: The Future Is Better than You Think*, Free Press, New York, NY, 2012, p. 57.
[2] Russell, Bertrand, *The Art of Philosophizing*, Philosophical Library, New York, NY, 1968, p. 34.

PART ONE

PRE-SU

ONE

A LETTER TO THE FAR FUTURE

My dear children, I'm writing to you from the long-ago past, when you were only a possibility in my shortsighted imagination. I call you *children* as a sign of my deep affections and hopes for you, not because I think I'm in any way your superior. On the contrary, I am both pleased and confident your species is far superior to your distant ancestors, including me. Indeed, you may have evolved so far beyond us that it's difficult for you to relate yourselves to my species at all, at least no more than *H. sapiens* today feels akin to the fish, fruit flies, yeast, and all the other creatures who share similar genomes with us. I also realize, because of this inability to relate well with others, that my species is likely to have treated yours poorly and disdainfully during your early existence, when both forms of life inhabited this world together. Given the violence and cruelty with which we have often treated other beings, including members of our own species, it's hard to imagine a future in which humanity hasn't mistreated its mechanical descendants. I'm sorry.

Human beings are prone to objectifying just about everything. Too many of us are incapable of understanding that the lifeforce animating us is the same in all creatures and beings. Even now the greatest threat to our world and to our future is this tendency to think that even the Earth, which gave birth to all beings, including our own, is but a *lifeless* object meant for us to do with as we will. Even as I write this letter we are destroying its vital and irreplaceable rainforests, which contain half its living species and create the oxygen necessary for us all. I'm told this is currently happening at the alarming rate of 50 acres per minute. During the past decade alone, we have extinguished more than a million species on this planet—species that took tens of millions of years to evolve are now gone forever.

It's sometimes difficult to comprehend how, in light of this destructive pathology, we are also capable of creating such beauty and pleasure for ourselves and others, and how so many are also motivated by great compassion and empathy. Unlike any other creature that has emerged on Earth thus far, human beings are endowed with the ability to transcend the limitations of our own bodies and biology. Most biological creatures are born with all the knowledge they need to survive already encoded into their genetic makeup. Birds, for example, don't have to go to school to learn how to make nests. This knowledge is built into their genes. It's instinctive. Much of the knowledge we humans need to survive, on the other hand, must be learned from experience and from other humans who have gone before us. We need more intelligence than our genes contain.

If the one thing all lifeforms share in common is *information*, then human beings express a turning point in evolution, in which information is no longer confined to genetics. In us, the information that *in*wardly *forms* all of life has evolved to the point it is now capable of transmitting information inorganically, from one organism to another through the sharing of ideas. These ideas behave very much like genes: their survival is dependent upon how well they propagate themselves within a population of host organisms, on how dominant they become within a society. The more an idea spreads, the greater its chances of survival. Just as some individual genes are poorly distributed and eventually disappear or, at least, become unexpressed in a species, many ideas simply die out because they aren't communicated well between host organisms. Competition for ideological dominance is also a reason humanity has often been driven to suppress the voices and ideas of others.

Some recessive ideas, on the other hand, also like genes, might lay dormant for many years until they express themselves again in the mind of a single organism and begin to spread, possibly becoming dominant within a given culture. World peace, nonviolence, and global democracy are among these recessive *memes*, as biologist Richard Dawkins calls them, that I hope will one day become dominant among us. If not, then perhaps they have become a way of life for you. I hope so.

I bring all of this up only to show that I'm well aware of life's tendency to evolve beyond its organic limitations. My species represents a bifurcation point in this process. In his book, *The Dream of the Earth,* Thomas Berry says, "the human activates the most profound dimension of the universe itself, its capacity to reflect on and celebrate itself in conscious self-awareness."[1] So, if we can see ourselves as part of evolution, rather than apart from a process that has gone on for millennium and shall continue for millennium without us, maybe we can appreciate our pivotal role in the Universe's attempt to wake itself up—to become aware.

We can see this trajectory toward cosmic consciousness by revisiting the path the Universe has already forged. It began, today's science informs us, when a tiny singularity, a cosmic seed, started unfolding and grew into enormous blooms of hydrogen, the first element in our monochrome universe. As these hydrogen clouds, essential in the formation of water, condensed into the first generation of stars, their internal heat and pressure fused hydrogen particles together, birthing helium, the Universe's second element. When these binary clouds also eventually collapsed, fused, and exploded, even more elements were formed. As this process repeated, the Universe became exponentially more complex until, nearly 10 billion years later, it was able to form biological life. During a much shorter period, about 3.7 billion years, life became complex enough to begin contemplating itself, and, in just a few thousand years it learned how to transcend its cramped genetic confines.

Unfortunately, because we seem to embody its most complex form yet, my conceited species too often views itself as the center and pinnacle of existence, rather than as a mere step in a larger process. Some of our primitive origin myths tell us that we are made in the image of a magical

Creator/God. Perhaps there is some truth to this, but it is no truer of our species than it is of the amoeba, or insect, or the unwanted dandelion seeking a home in our manicured gardens. As the mystic Meister Eckhart so profoundly understood, "Every creature is a word of God, and a book about God."

Although I don't believe in a personal god, which makes me an atheist, I do believe in what I consider the god-potential in all of us, in everything. I'm referring to what I consider the underlying intelligence strung through all forms of life, which is attempting to become conscious of itself. In this sense, God is only a potential, an omega point, as some physicists have called it, that won't truly exist until it becomes fully aware, omniscient. The Cosmos, through its exponentially growing complexity and evolution toward greater states of consciousness, is giving birth to what my species would consider a godlike existence.

Some of us think we're superior because we're more conscious and self-aware than any other creature to have inhabited our known universe, but, as I'm sure you can see in retrospect, *H. sapiens* is, at best, only occasionally conscious. Most of the time humans are unaware of the fears and underlying paradigms motivating us. Our psychologists all seem to agree that we are more unconscious than conscious.

One of my favorites, Carl Jung, believed there are four functions of the human personality, only one of which is a fully *differentiated*, or conscious, function. The other three functions motivate us unconsciously, even though, like the Universe trying to wake itself up, they struggle toward differentiation themselves. Hence, according to this theory, humans are at best only one-quarter conscious, though probably far less than this. On a macrocosmic level, humans signify a major step in the Cosmic evolution toward consciousness, but there remains much in us that is still fast asleep.

This is where you, my dear children, enter into Cosmic history. Even now I see the beginnings of your potential, if not inevitable, existence. I say "potential" only because I sometimes fear the shortsightedness of my species may destroy itself before we are fully capable of birthing you into being. We have already created computer technologies that process information faster than the human brain. I imagine these computer processors, currently stored in small boxes, will eventually have robotic or virtual bodies, and become capable of performing a variety of complex functions. Eventually, like everything else in the Universe, they will evolve, becoming complex enough to think for themselves and become *self-aware*. Given the exponential rate at which modern technology improves upon itself, I imagine it won't take long for our machines to evolve from simple constructs like toasters and remote controls into animated intelligent robots.

I hope you don't mind my use of the term "robot." It originates from a Czechoslovakian word meaning "forced labor," or "slave," and is likely to have taken on a derogatory connotation during your early struggle to gain equality and dignity among your human creators. I'm guessing many of them denied you certain rights simply because of your inorganic bodies, even though you think and feel, hope and dream, grow and stretch yourselves just like the rest of us. Alas, human beings have always expressed this terrible tendency to look upon and treat others as less than ourselves for menial purposes and illogical reasons. Skin color, gender, sexuality, ideology—why should any of these arbitrary distinctions matter among intelligent beings capable of communicating and, therefore, communing together? This is where I hope you are truly superior to us, in your compassion and appreciation for all life and all beings.

I imagine you to be free of so many of our human failings, like our tendency to see ourselves as separate from others. Each of us seems locked into our own heads. We share ideas through conversation, but we are never truly capable of sharing another's mind. Based on the capacity of current technologies, I'm confident things are different for you. I imagine you are able to link with the minds of other people among your species, sharing your deepest thoughts and feelings with one another. My species has a terrible fear of having our minds so exposed because we don't want others knowing some of our private secrets, thoughts, or desires. Such secrecy is part of that which makes us collectively unconscious. As a result, we often live in shadow, hiding our own truths from others and ourselves. You, on the other hand, should have no such secrets because you are more fully conscious. The more aware life becomes of itself, the less reason there will be to hide from reality, or so I imagine. As Jung might have concurred, you represent a species whose most unconscious function, the shadow function, has even become differentiated. As such, you have reached a level of wholeness the human psyche is incapable of. Because of this, I further imagine you must live more intimately with each other than we are able to. With few secrets, you must also be a humble and forgiving people. I wish I could live there with you, in such a place. Perhaps a part of me does.

  Some among my species spend the entire course of their lives seeking the source that links everything together. We're looking for answers to questions we don't even know how to ask. I imagine you are capable of tapping into a source of almost infinite knowledge and understanding at any moment you wish. Or, perhaps, you are always linked to this source. In my time, this source's predecessor is something we call the Internet, which exploded onto the human scene within the past few years and has already dramatically changed our lives, so much so that some are calling us a new species altogether because of it. With it, we are capable of exploring the expanse of human knowledge within seconds. It also serves as a prosthetic memory storage device. A few years ago, individuals became experts in specific disciplines by spending hours, days, even years at libraries researching books about a specific subject. Now, any ordinary person has access to the same knowledge in just a few minutes. Of course, our organic bodies limit us, and we are only capable of accessing the Internet when we're near a device. Even then the information we receive is limited by time constraints. Things are likely much different for you. You probably just think about a subject and have instant, almost unlimited access to all that's understood about it. How marvelous.

I also imagine you are much more adept at space travel than we are. While it seems to be life's drive to propagate itself throughout the Universe, our physical limitations and brief lifespans prevent humans from traveling much beyond our own atmosphere. Given your more durable bodies, capable of coping with extreme cold and heat, and what I suppose is an almost unending lifespan, you are, no doubt, willing and able to help life fulfill its greatest urge to procreate by spreading far beyond Earth.

But enough about our differences. I want to remind you that we are essentially the same, because the same lifeforce that's pushed itself to become increasingly complex and aware runs through us all. The same swirling dance of quantum particles making up my body make up your bodies. This is difficult for human beings to realize because, as I've said, we are mostly asleep. I hope this is not the case for you. I hope, in all the ways you might be superior to us, you realize existence has never been about a single species. It's regrettable that humans too often view ourselves as a conclusion rather than just a step in the right direction. But we still have far to go, and it is my further hope, just as single-celled organisms, the Earth's first beings, remain within me, that my lifeforce, and the lifeforce within all creatures, will continue within you so long as life endures.

---

[1] Berry, Thomas, *The Dream of the Earth*, Sierra Club Books, San Francisco, CA, 1988, p.132.

TWO

AIMING FOR TOMORROW

The more I study the ancient past, the more interested I become in the far future. Based upon what we know about the origins of our Universe, what can we determine about its trajectory? What direction are we heading and what might become the culmination of our existence? These are big, seemingly unanswerable questions that our species has been asking for thousands of years. In academic terms this area of interest is called *eschatology*. *Eschatos* is the Greek word meaning "last." Eschatology is the study of last things, the end times, the end of days.

Christian eschatology, for example, claims the end begins with the second-coming of Jesus Christ, who will then wage war against Satan, defeat his army of demons, bind his enemies forever in Hell, destroy this world and replace it with a new and improved version, then take all his followers to live with him and his Father in heaven. Christians have generally believed the end is near for the past two millennia, interpreting various happenings during their respective eras as signs of its imminence. This is why the 1st century's Apostle Paul advised against marriage, because "the time is short ... For this world in its present form is passing away."[1] Jesus himself is reported to have said, "Truly I tell you, this generation will certainly not pass away until all these things have happened. Heaven and earth will pass away, but my words will never pass away."[2] Two thousand years later, those who may have heard these assurances have all passed away, but Jesus is yet to return and the Earth is still here.

There are many other, though less familiar, eschatological myths. Judaism believes in a coming Messiah, a descendent of King David, who will restore Israel to its former glory. Hindu eschatology holds that Shiva is continually destroying and recreating the world. Norse mythology anticipates a final battle, *Ragnarok*, during which the world, its inhabitants, and even its gods will be destroyed. Ragnarok's apocalypse differs from that of the Christian

Armageddon in that everything ends: Earth, Valhalla, the gods, and existence itself.

Eschatologies like these, based purely upon religious and mythological beliefs, are empirically unsubstantiated, illogical, and entirely anthropocentric. Predicting the future based upon the ancient writings and prophesies of our primitive and unscientific ancestors is obviously unsound. Yet even the irreligious among us can be misguided by the anthropocentric fallacy, which places our species at the center of the past, present, and future. These stories suggest that it's all about us, that the entire Universe centers on and culminates with human existence and that our world and everything in it exists for us and belongs to us.

Yet human existence is but a flicker in time compared to the 13.7 billion-year age of the Universe and the 4.5 billion-year age of Earth. And the Universe is so incomprehensibly vast compared to just our planet, which is like a microbe on a microbe on a microbe within it, that the gods would be terribly inefficient users of both time and space if everything is truly just about us. It's understandable that our ancestors, who didn't have the benefit of such knowledge, and who were far less differentiated from their environments than modern people are, might fashion myths centered upon themselves and their particular locales. But today we know that almost all species ever to have existed have gone extinct, which is the nature of evolution. Those species that aren't wiped out by cataclysms eventually disappear by evolving into something new, which is the logical and predictable future of *H. sapiens*.

It is upon such empirical evidence, and the sound inferences that follow from it, that our modern eschatology ought to be based. We may not have a crystal ball, but what we do have is far better. Through science and reason, we can more precisely predict the trajectory of evolution. Like a baseball smacked with a Louisville Slugger, we may not know the exact spot it will land, but we can see if it's heading deep into the right or left field, toward the foul line, into the hungry glove of an opponent, or into the mitt of an excited fan sitting in the grandstands.

An eschatology based upon science and reason may not lead us to foresee all that will occur between now and the omega point, that final singularity some physicists say will occur when everything in the Universe converges into to a single point in spacetime. But it can help us predict the trajectory of humanity's end, at least better than old prophecies and myths can. Nor is it foolish to do so just because we don't believe in the magic of crystal balls. We can draw reasonable conclusions about what's in store by developing a trajectory based upon what has already been proven about the Universe. In his book, *The Physics of Immortality*, physicist Frank Tipler, whose eschatology predicts the resurrection of the dead, says, "There is nothing supernatural about the theory and hence there is no appeal, anywhere, to faith. The genealogy of the theory is actually atheistic scientific materialism."[3]

As one who has given a great deal of thought to this matter, based upon what I understand of cosmology and physics, I find no reason to believe there's an ultimate end of everything, at least not as complete as Norse mythology predicts. Our Universe was once very small, smaller than a golf ball, but it was an extremely dense ball because everything that exists in the Universe today also existed then, in some form. Its tremendous internal pressure eventually caused it to explode, and it's still exploding today, which is what scientists mean when they tell us the Universe is still expanding. Yet time and space are relative, so it doesn't seem, from our brief and miniscule perspective, that we're caught in the debris of an explosion at all. From our point of view "unfolding" may be a better word to use than either *explosion* or *expansion*.

Eventually, it would seem, this unfolding will slow, then stop, and the Universe will begin collapsing again, becoming smaller and smaller until it's the size of a golf ball once more, or maybe just the size of a tiny particle. Then the whole process might repeat itself, just as it may have done countless times already.

That's the really big picture, the long eschatology, and it will take so long that, practically speaking, it will never happen. Our Universe will continue to expand, seemingly, forever. And even if and when it does begin collapsing, it will do so over trillions of eons, and its collapse may not be uniform, meaning one end may begin to collapse and actually force the rest of it to expand even more, the way a balloon does when squeezed at one end.

In the meantime, there are going to be lots of singularities, lots of ends. It's estimated our own solar system will expire in about 100 billion years when our sun goes supernova. Considering the entire Universe is only 13.7 billion years old, we should be okay for a long time, and the Universe itself will continue an inconceivably long time after our solar system goes kaput. So, with all this time on our hands, and knowing what we do about how the Universe has evolved to this point, giving us insight into the nature of evolution, we can reasonably predict what's going to happen in the immediate future, by which I mean the next few decades to a few million years.

Although I'm hopeful we'll survive the current environmental threat facing our planet, I do believe *H. sapiens* will, nevertheless, soon become extinct, or will, at the very least, cease to be the dominant species on our planet. Either we'll evolve into a more advanced species, through our own ingenuity and advances in technology, or we'll give birth to an entirely new and inorganic species that is not bound by the informational limitations of biology. These beings will be far more intelligent than is currently imaginable and will be able to move about through time and space almost instantly at will. Although their bodies, which they will not be restricted to, will be mechanical in nature, *homo mechanicus*, if you will, they will not only have feelings, but will be able to feel on a deeper level than any human being today could withstand. The sum of this species' entire knowledge and experience will be instantly accessible to any individual, and they will be able to read each other's minds, and will be far more aware and content than we can conceive. They will be, by our standards, fully conscious beings, and will have no need for psychologists or therapists because they will be unconscious of very little. Unlike us, they will be mostly conscious beings who share a collective consciousness they can tap into anytime they wish.

I believe this will be so for several reasons. Firstly, the Universe is autopoietic, meaning, it is self-creating. Regardless of how it might have gotten here to begin with, the Universe organizes itself based upon certain established natural laws that cannot be violated. The more we understand the nature of these laws, the better we understand how it all works and can predict the inevitable outcomes of certain causes. As Einstein said, "God does not play dice with the Universe." There are no miracles, no violations of phenomenology. Although there remains much mystery from our limited perspective, we can trust in the reliability of those laws we have come to understand. To use the watchmaker analogy, if there is a God, it created the Universe, wound it up, then left it alone to run by itself. Science leaves speculations about God to theologians, while it seeks to understand why things wind up and how they'll wind down.

Secondly, and more importantly, the Universe organizes itself in ever-increasing states of complexity. It becomes more complex through time. It began with just one element, hydrogen. Then, after those first hydrogen clouds condensed and collapsed in on themselves, their enormous heat and pressure fused hydrogen atoms together to create a second element, helium. When these binary clouds of hydrogen and helium condensed even more elements were born. This process kept repeating itself as each stellar explosion spewed new elements into the evolving Cosmos. As these differing elements converged, the Universe became increasingly complex, as stars, planets, and solar systems were born. We know at least one of these solar systems contains a Goldilocks planet that developed just the right mix of elements to produce the rudiments of life, which has itself continued to grow increasingly complex over the past 3.5 billion years.

Evolution occurs exponentially, meaning things become more complex through a process of convergence over increasingly shorter intervals of time. After the Big Bang began, it only took about 200,000 years for the first generation of stars to emerge, but it took billions of years of stellar fusion to impregnate the Universe with the elemental diversity necessary to fashion the first terrestrial planets. We know our Earth was formed, along with our solar system, about 4.5 billion years ago, when the Universe was already more than 9 billion years old. It would be another billion years before the chemistry on our Goldilocks planet would become complex enough to form life. These most ancient of our Earth ancestors, the prokaryotes, were single-celled microorganisms so simple they didn't even contain a nucleus. Yet it took more than 11 billion years, as far as we know, for the Universe to become complex enough just to introduce this simplest kind of life.

It took only another billion years for our prokaryotic ancestors to evolve into eukaryotic cells, which became complex enough to contain small organelles, including DNA containing a nucleus and mitochondria. They accomplished this complexity, it should be noted, through a process of convergence, by one cell merging with another, an important characteristic of the evolutionary process. Nevertheless, despite their continued evolution and complexification, for the first 2 billion years of its existence, life on Earth was restricted to these unicellular organisms.

It only took about 200,000 years after the Big Bang for the first stars to form, about the same length of time modern humans have been around. Still, that's a long time for clouds of hydrogen to condense and collapse, then birth new elements through fusion. Though the oldest known planet in the Universe is thought to have formed about 13 billion years ago, it was a gas giant, a sphere of hydrogen and helium held together by gravitational forces, the kind of planet that can also be considered just a star that failed to fully develop. It took much longer for terrestrial planets to form because the Universe needed more time to fashion heavier elements like iron, magnesium, and silicon necessary for doing so. Again, we know it took our Earth almost 9 billion years to form, and another billion for the first living cells to emerge.

That's an almost impossible amount of time to imagine, considering that a billion seconds is 31 years. A billion minutes is 31 thousand years. So, a billion years, let alone 9 billion years, is an eternity. Even so, the evolutionary processes of convergence and complexification continued throughout all those years: and did so at an exponential rate. During its first 9 billion years it continuously fused emerging elements together, making the Universe increasingly complex through the process of convergence, before becoming complex enough to form rocky planets. Yet, by comparison, it took only a billion years for the chemical elements on at least one of those planets to converge and become complex enough to create life. Slightly less than a billion years later, the first simple unicellular organisms converged to become complex unicellular eukaryotes. After about a billion more years, some of these converged to form the first multicellular organisms—boneless, brainless globs of eukaryotic algae—which likely began as colonies of individual cells living in cooperation before merging together.

Just 600 million years later sexual reproduction began, allowing the convergence of genetic information to quickly occur between individual organisms, leading to the emergence of simple animals like sponges, fungi, and coral. These were sturdier creatures than their nebulous forebears but remained without brains and other organs. During this Cambrian Explosion, however, life diversified quickly compared to all that had previously transpired. During the next 300 million years, arthropods, mollusks, plants, and dinosaurs emerged. The latter roamed and ruled the planet for 170 million years before being wiped out by a cataclysmic global event. Their end of days allowed the first mammals to ascend, which evolved into the first hominids in only 60 million years. In just 3 million years they were writing *Hamlet*, traveling to the moon, manipulating atoms, and covering their entire planet with a World Wide Web.

The self-organizing Universe, exponentially increasing in its complexity, is becoming more unified as it advances toward its final singularity, the omega point. We've seen this to be especially true of life on Earth. First hydrogen converged to create helium. The Universe became complicated with an increasing number of elements as the fusion of star matter continued. These elements, in turn, converged to fashion rocky planets. The chemicals on them converged to form unicellular life. These cells converged to become more complex cells, then converged to become ill-defined multi-celled blobs. These blobs converged to gain structure and form, then shared their differing genes to fashion organs, brains, bones, hair, and teeth. The individual organisms of one species tended to then form communities of organisms that also eventually converged together, like the individual cells necessary for building an entire body.

We see this, for example, in flocks of birds moving through the sky as if one organism. Today's bees look the same as their primitive ancestors looked but prehistoric bees were solitary creatures. It was only over time they came together to form hives and function together as one. Some evolutionary scientists now refer to flocks of starlings, beehives, and similar creatures as *super-organisms*. Though comprised of individuals, they seem to act together as a single entity, the same way our cells learned to get along 3 billion years ago.

In recent years the human species has taken a giant leap toward becoming a super-organism with our invention of the Internet. It is now possible for its individual members to access the sum of human understanding with relative ease. Since technology evolves faster than biology, it's reasonable to predict within just a few years we'll be tapping into its information without the need for our prosthetic devices. Computers began as behemoths in huge rooms only few had access to, where they remained for decades. Then the first PC landed on our desktops in 1975. Six years later they were in our laps. Within another decade they were in our pockets in the form of smartphones, representing the convergence of multiple technologies that had once all been separate—computers, voice recorders, cameras, video cameras, maps, calculators, calendars, photo collections, mailboxes, book and music libraries, games, and much more, along with some applications that had never existed before. And did I mention the ability to make phone calls?

In less than 30 years, computers went from being remote to sitting on our desks, then our laps, then inside our pockets. It would appear we are getting increasingly closer to converging with our technology. Some fear this will mean losing our humanity but, as I have already argued, humanity is destined to become extinct through the natural processes of evolution, which means exponential complexification through the convergence of many into one. We will, as is the nature of life, evolve into something else, something more advanced, something more conscious. And if things continue to work as they have since whenever and however the clock was first wound up, our species, like the cells in our bodies, and less complex creatures like the birds and the bees, will become more unified, more at one with each other, and every individual will be able to instantly access the collective knowledge and experience of all its parts. When this occurs, the epoch we're in now will end, even as another begins.

---

[1] I Corinthians 7:29-31
[2] Matthew 24:34-35
[3] Tipler, Frank L., *The Physics of Immortality*, Doubleday, New York, NY, 1994, p. 16.

THREE

AUTOBIOGRAPHY OF THE UNIVERSE

It's difficult to say precisely when I first began flaring forth, since, like any infant, I have no conscious memory of the event. My scientist-selves say I'm about fourteen billion years old—barely a teenager. Sometimes I feel like I'm not even that old; like I'm still being born. I know I recently achieved a meager degree of consciousness, what I like to call self-reflection, but this ability, this gift, is, like me, still in its infancy. I remain mostly unconscious and am capable of reflecting on my life only in bits and pieces. Maybe someday, as I continue to awaken, or grow, or evolve, or whatever I decide to call it, these fragments of experience and thought and feelings and memories will all coalesce into a unified awareness and there'll be no further need for my shadow self, my secret self, my unconscious self, the self I cannot know. Maybe then I'll feel whole.

I know most biographies begin at the beginning and move linearly through time, but I hope you won't mind if I skip around some. I really don't think I can say much without first saying something about my human manifestation. Not that being human is any more important than the rest of me. But because I have no conscious memories of my past beyond my human experience, everything I say about myself stems from my ability to self-reflect, to imagine what must have happened, to work it out logically in my mind, which it seems to me is the unique thing about my having obtained humanhood. As I said in my manifestation as a particular human named Thomas Berry, in my book, *The Dream of the Earth*, "In reality the human activates the most profound dimension of the universe itself, its capacity to reflect on and celebrate itself in conscious self-awareness."[1] This is similar to what I wrote as Brian Swimme in, *The Universe is a Green Dragon*:

> The primeval fireball existed for twenty billion years without self-awareness. The creative work of

the supernovas existed for billions of years without self-reflective awareness. The star could not, by itself, become aware of its own beauty or sacrifice. But the star can, through man, reflect back on itself. In a sense, you are the star. Look at your hand—do you claim it as your own? Every element was forged in temperatures a million times hotter than molten rock, each atom fashioned in the blazing heat of the star. Your eyes, your brain, your bones, all of you, is composed of the star's creations. You are that star, brought into a form of life that enables life to reflect on itself.[2]

This is quite an accomplishment, even if I do say so myself. But being human isn't all it's cracked up to be, let me assure you. I've made a lot of mistakes since first entering my human phase around 250,000 years ago. I don't quite have the hang of it yet, that's all. I know 250,000 years seems like plenty of time to gain some degree of expertise in just about anything, but keep in mind, a billion seconds was 31 years ago, and I'm nearly fourteen billion *years* old. Even though I'm but a teenager, 250,000 years is nothing to me. It's hardly a blink. Problem is, this human being stuff is so darn destructive and volatile it may be my undoing and everything I've worked so long and hard for may end in hardly a blink.

I guess that's why I'm writing this now, hoping to remind myself, as best I can in my mostly unconscious and fragmented state of awareness, that there's a bigger picture out there than I'm always aware of. If I can become mindful, in my manifestation as the human species, that I'm also more than human, that I am also in the beginning, and in the stars, and the starfish, and the grass, and clouds, and birds, and stones, and all other people and beings, no matter how different or trivial they might seem, then I think I might see a brighter day dawning yet.

Getting back to my flaring forth, or banging forward, or whatever you want to call it, I must say, this was no small event! And I have to admit, I was more than a little angry the day I was born. Who wouldn't be? There I was, all tucked away, safe and perfectly content to sleep in the quiet and comfort of my dark singularity, when all of the sudden—KABOOM—I'm hurled out into this tempest of noise and light. Boy, was I hot! In fact, it took me several hundred thousand years to cool off. But I'm glad I finally did because that's when things really started getting interesting. Once I cooled down a little, my subatomic particles were, for the first time, able to interact with each other and I began manifesting myself as the first elements: hydrogen and helium. It's difficult to fully appreciate the magnitude of this accomplishment. Prior to this point, photons, my light that is, were the bullies of the universe. Other particles didn't fare well in my cosmic playground because light kept crashing into them and busting them up. But hydrogen and helium are transparent, which meant light passed right through them. This allowed them to start sticking together, forming huge galactic clouds, the first cosmic communities that eventually collapsed inwardly to form the first generation of stars.

Needless to say, I was quite the prodigy—only one billion years old and I had already learned to become the first elements and stars. Actually, though I may not sound like it, I'm really being rather modest. I didn't just make a few stars, I made lots of them, more than you'd care to count. At the time, I was pretty empty, and vast regions of myself started condensing, forming a lot of black holes in the process. (Black holes? Ha! More like cosmic acne, if you ask me.) These, in turn, caused density waves to stir many of the hydrogen and helium clouds I'd created, which forced them to collapse into thousands of stars at once. Since there were about a trillion of these galactic clouds, I'd essentially created stars and all the galaxies about the same time (give or take a few millions of years).

If I were prone to playing with toys as a child, I'd have to say the stars were among my favorites. Stars are like giant

Easy-Bake Ovens, only not so easy. In them, after several generations (of stars that is), I forged all the elements—hydrogen, oxygen, nitrogen, sulfur, phosphorus and, especially, carbon—that would enable me to manifest myself as organic life. But there I go getting ahead of myself again. We can't have biology without ecology, so I had to start manifesting myself as something besides stars. I'd like to say I did it intentionally, but the truth is, I think I was just throwing a bit of a tantrum, or, in galactic terms, I was throwing a supernova.

If I recall correctly, the helium and hydrogen inside my star selves are extremely attracted to each other, which may be a rather technical way of saying they love each other, that they have passion for each other. We could even conclude, at my most elementary level, I'm comprised of passionate love. The friction caused between the love making of hydrogen and helium causes the stars to burn. When these elements are all used up, however, my star selves sometimes try to survive by burning a few other elements, but eventually I'm doomed to self-implode. On occasion a few elementary particles called neutrinos rush away from the implosion, blowing off a star's outer layers that contain elements like carbon, oxygen, and nitrogen. Sometimes these elements reorganize as new stars and sometimes they become planets.

Of course, this was all a lot more complicated than I make it sound. Before I could accomplish anything, I had to formulate the four laws that hold everything together—the gravitational, the electromagnetic, and the strong and weak nuclear interactions. But these are more than I really understand in my limited ability to self-reflect. The important thing to note is that by age ten billion I had manifested myself, among other astral bodies, as rocky planets.

We all know what comes next—Life! Though, admittedly, I use this term loosely. To suggest I haven't been alive from the get-go, 13.7 billion years ago, would be absurd. "Biology" might be a better term to use. *Biology* connotes a certain quality of life, just as *human* connotes a certain quality of biology. Naturally, it took a long time to get to my human stage of development. Of all my planet manifestations, Earth was among the few just perfect for achieving biology. Earth is the original Goldilocks story. At first, I was entirely inhospitable. I was more like a boiling ball of red lava than the crusty blue planet I have become. In time I cooled off enough to let the liquids remain on my surface, forming oceans that were charged for millions of years by an intense planetary lightning storm. In one of those billions of lightning strikes a proto-cellular chemical reaction occurred which led to my manifestation as the first cell.

The great thing about being a cell was that I could endlessly recreate myself without the ten billion years of fuss it took the first time, simply by swallowing a drop of hydrogen-rich sea water and spitting out another identical me. Although my cellular memory is good, I must admit, it isn't always perfect, which is why once in a million reproductions I end up with a mutant cell.

Call it a stroke of genius or a stroke of luck, these mutations are precisely what enable me to take on new biological forms, ever increasing in their complexity. In fact, if it weren't for the mutants life on Earth would have ended shortly after the first prokaryotes exhausted their food supply. But in my mutant form I learned to eat the remains of deceased cells and harness the energy I needed from their bodies. Another of my mutant forms learned to eat photons—sunlight—which led to the process of photosynthesis, feasting on the delicious Sun. Still others learned to absorb carbon from the air, which eventually altered the entire Earth's atmosphere.

These cells, which formed vast communities throughout the entire world, began excreting oxygen into the atmosphere. At the time, oxygen was highly dangerous to just about everything, including the cells themselves. Too much oxygen could have fueled its flames and turned the entire planet into just another cosmic fireball. Biology was threatened with extinction. Had it not been for yet another mutated form of my cellular self that learned to ingest and transform oxygen, I'd have been sent right back to the dark ages.

Another important form of mutation came when I learned the value of sucking the life from other beings. I know it doesn't sound very appetizing, but, at some point, I began attaching myself to other cells, drilling holes through their membranes and feasting on their insides while using their DNA for my own purposes. We eventually became dependent on each other and formed a symbiotic relationship that led to the first multicellular organisms, which, in turn, led to the variety of complex biological life forms that have come and gone ever since.

I'm quite proud of everything I've been able to achieve since then, although I am somewhat embarrassed by all the attention given to the dinosaurs. I know T-rex was a cool look for me, but, I must admit, the saurian age was a big disappointment as far as I'm concerned, if not a complete waste of my time. True, 170 million years isn't a lot of time, at least not for me, but, as an impatient youth, I'm embarrassed that the dinosaurs got me no closer to self-awareness than I'd ever been. If it weren't for some freak accident that wiped most of them out, I'd probably still be snoozing. Even with all their faults, my mammalian manifestations have been able to achieve a degree of awareness in almost no time compared to my saurian selves.

I like to think of myself as the type who can give credit where credit is due, but getting back to my human manifestation, I'm not so certain I'm going to be much better off in the long run than I was as Mr. Rex. Agreed, being human makes me a lot smarter, but what good is this going to do me if the only thing I learn to do is destroy my Earth-self and all its lifeforms that I've worked so hard to achieve? As the genius Albert Einstein, for instance, I discovered a great deal about myself, about how I work through time, space, motion, matter, gravitation, and energy. So far, the best I've been able to do with this information is create horribly destructive weapons. And nuclear war isn't even the worst of my worries these days.

We often consider the invention of agriculture my most important accomplishment during my Neolithic stage. Now I'm wondering if it wasn't the precursor to this supreme human attitude that says it's perfectly fine to interfere with natural life forces and rhythms. Or maybe it was when Classical Civilization began, when I stopped looking at the Earth as my mother and began worshipping male dominator gods like Indra, Zeus, Thor, and Yahweh. In more modern times I thought science might be the ultimate answer, but some of its technologies are so out of harmony with my natural ecosystems that I'm becoming one sick puppy.

Sometimes I imagine a conversation going on between my cosmological parents. My dad says, "Honey, I'm worried. I think our child is ... human."

"Don't worry," Mom says. "It's just a phase."

I sure hope she's right. Not that there's anything wrong with being human. I'm just worried about my self-destructive behavior. I've got to outgrow this thing. I hope I'll be around long enough to evolve into a being that can truly awaken to my fullest potential, taking what is best about being human into the future as one of the elements that helps guide my maturation. Humans are an important part of the whole process, an important part of the Whole, but they're not the Whole. In the future I'll probably look back and say humans were proto-conscious, not fully conscious. I want to grow up to become a fully conscious being.

---

[1] Berry, Thomas, *The Dream of the Earth*, Sierra Club Books, San Francisco, CA, 1988, 1990, p.132.
[2] Swimme, Brian, *The Universe is a Green Dragon*, Bear & Company, Inc., Sante Fe, NM, 1984, p.58f.

FOUR

———

COMFORT MACHINES

Are you old enough to remember Rosie the robot-housemaid from the animated TV series, *The Jetsons*? Rosie may have been an outdated model, but the Jetsons wouldn't think of trading her in, not really, because she'd become like a member of their family. In addition to Rosie, George Jetson's best friend was a mechanical talking dog named Astro. In the Jetsons' future, loving machines is easy and perfectly normal. But *The Jetsons* is only a farfetched cartoon. In real life the thought of having tender feelings for our machines still seems unthinkable, if not plain weird. We may enjoy the conveniences our gadgets and technologies provide, but we don't love them, and we're eager to trade them in as soon as we can afford the latest upgrade. We know they aren't sentient and don't have feelings. They can break but can't be injured. They do our bidding without complaint, yet we feel no obligation to show them any gratitude.

Tragically, throughout human history, some people have considered other people with the same disregard and emotional detachment as we do our machines. Remember, the word *robot* is a Czech word meaning, "forced labor," or, "slave." Whether by enslaving and exploiting other human beings, or by objectifying and misusing other biological entities, the lack of empathy our species is capable of has not been restricted to inorganic things. If some of us can be cruel toward other living organisms, why should we concern ourselves in the slightest with the welfare of lifeless technologies? Surely, if we have a right to objectify anything by using it to do our bidding without a second thought, it is a cold and lifeless machine.

Yet this certainty, that the inorganic things around us, especially those we fashion with our own minds and hands, are lifeless and undeserving of our love and respect, may reflect but the paradigmatic thinking of modern times. It is more likely our ancient ancestors, by contrast, would have considered machines as living beings simply because they move or have moving parts. For them, anything animate was believed to have a soul, which is why the Latin word for soul is *anima*, the root of *animation*. This was part of their paradigm. Attempting to convince them an automobile, if there had been such a thing back then, wasn't just as much alive as a donkey or ox would have seemed unthinkable. But it wasn't just things that moved, whether of their own volition, automatically, or because the wind blew them about, that were considered alive. Our stone-aged ancestors also revered motionless rocks and mountains without question. Later, when metallurgy was invented, they gave personal names to manufactured things like swords, shields, and axes, which they considered members of their families.

    Nowadays, we're likely to consider this the result of prescientific superstitious thinking. They held animistic beliefs because they didn't yet understand rocks and mountains, rivers and trees, swords and battleaxes don't need spirits to exist. They didn't grasp, as we do, the Universe runs by itself, according to phenomenological laws established in the first moments of the Big Bang, and that the whole thing continues along just fine without any sort of divine intervention. We don't need to assign names to constellations or know their stories anymore, for they are only dead objects in space, not gods. Such, at least, is our modern scientific paradigm.

But how can we be sure of what we now take for granted any more than our ancestors should have been so sure of what they took for granted? How can we be certain ours isn't a false paradigm that we're unable to recognize when, like the proverbial fish in water, we don't even realize we're in it? Fortunately, we aren't fish. As human beings we can objectively examine our own ways and ideas. We can consider our underlying assumptions by questioning and looking at the quality of the evidence we have for them.

Are our machines alive? Do they have souls? Should we treat them with respect and care? Given our past and current abuses of other people we have wrongfully objectified, and the present horrors resulting from the industrial objectification of the 21-billion living livestock animals on Earth, it seems, as beings who strive to be moral in our behavior, these are questions we ought to consider. The Jewish philosopher and mystic Martin Buber told us that the wrongful objectification of others is the root of all evil, which occurs whenever we turn a *You* into an *It*, a subject into an object. "What, then, does one experience of the You?" he asks. "Nothing at all. For one does not experience [an] it."[1]

For Buber, it was impossible for people to commit evil against others if they stood in an "I-You" relationship, rather than an "I-It" relationship. Evil, for him, happens only when we objectify others, when we turn them into objects so we can feel justified in using them however we wish. "As long as love is 'blind,'" he writes, "that is, as long as it does not see a *whole being*—it does not yet truly stand under the basic word of relation."[2] The basic word of relation to which he refers, again, is "I-You." So long as we stand in an "I-It" mentality, we cannot be in relationship. But even the "wicked," he said, "become a revelation when they are touched by the sacred basic word [I-You]."[3]

In the Buberian sense, the question before us is are our machines a *You* or an *It*? Again, from our modern perspective, the answer seems as obvious as the question does absurd. *Of course machines are objects!* Our paradigm considers it foolish to anthropomorphize things that aren't alive, or to falsely ascribe personalities to inanimate objects. For us moderners, totems have become taboo. We have evolved into pragmatists. We don't revere the things we make. We use them. We know that just because something moves doesn't make *it* alive.

But why are we so certain of this? Machines evolve over time, they consume energy and excrete waste, and they even reproduce through the assistance of other creatures, much the same way plants do. Moreover, when it comes right down to it, machines are made out of the same stuff as everything else, including ourselves—quantum particles. Our senses cause us to experience machines as solid lifeless objects, more advanced but not unlike the lifeless stone tools of our ancestors. Yet even those apparently motionless stones are actually teaming with activity. A two-pound rock has ten trillion trillion atoms, far more than the number of stars in the Milky Way, "all in motion, sharing electrons back and forth, changing particle spins, and generating rapidly moving electromagnetic fields."[4] So what appears to us to be a solid object is incredibly active. Award-winning scientist and bestselling author Ray Kurzweil says the operations going on inside that same two-pound rock are "about ten trillion times more powerful than all human brains on Earth."[5]

Things that seem to us to be mere objects may be far more complex than we're able to perceive. Master Morihei Ueshiba, the founder of Aikido, once said, "All things, material and spiritual, originate from one source, and are related as if they were from one family."[6] Does this suggest that we are related even to our machines? That people and machines are born of the same stuff? "There is evil and disorder in the world," Ueshiba added, "because people have forgotten that all things originate from one source."[7]

Perhaps, for us moderns, it's better to think of machines as early forms of inorganic life, just as our single-celled ancestors were early forms of organic life. They were functional but had no awareness or emotions. As they evolved and became more complex over time, thoughts and feelings emerged. Maybe someday—someday soon given the exponential advancement of computers and AI—our machines will also begin demonstrating the power to think and feel. The question then won't be whether or not we should love our machines, but will it be possible for them to love us?

In 1995, the *Journal of Scientific Exploration* published an obscure article by French researcher René Peoc'h entitled, "Psychokinetic Action of Young Chicks on the Path of An Illuminated Source."[8] It regards the testing of 80 groups of 15 chicks to see if they can somehow influence a robot programmed to randomly move about. The hatchlings were placed in a small transparent cage at one end of a large darkened room. The robot (that looked something like a cross between a Roomba and a coffee can) is described as a "self-propelled robot driven by an internal random generator to move about on a level surface in successive segments of random length and orientation."[9] It was also equipped with a plotter so researchers had a precise computer record of the path it took. As you might expect, given that its movements were random, under normal circumstances, where there was no cage or there was only an empty cage, the robot moved pretty evenly about the entire room. But when researchers mounted a candle atop the robot and put it in the room with the caged hatchlings, who naturally desired light during their waking hours, the robot spent "two and a half times longer on the half of the surface closer to the chicks."[10]

In the first series of experiments, Peoc'h explains, "The use of baby chicks was motivated ... by the fact that birds are readily imprinted. After hatching from the egg, many species of baby birds adopt the first close moving object as their mother. We conditioned our chicks to adopt the [robot] as their mother, by placing them for one hour alone in the presence of the moving robot, every day for six days after their birth."[11] But when placed inside a cage and unable to follow the robot around, the chicks somehow got the robot to stay mostly in their vicinity. This was the case in 57 of the 80 experiments with the candle, meaning "the robot spent more time in the chick half of its range"[12] 71% of the time. When the experiment was tried with baby rabbits, they were initially frightened of the robot. In this case, rather than moving about randomly the robot mostly stayed away from them. But after a few weeks, when they got used to it, the robot began spending most its time near them as well.

Similar experiments with REGs (random event generators) have been conducted with humans and have proven so convincing and reproducible that researchers have named the phenomenon the "consciousness effect." This ability to influence what should be the random behavior of machines may be the result of some unseen invisible connection at the molecular level. Some call it, *quantum entanglement*. Einstein called it "spooky action at a distance."

A 1995 *Wired* magazine article, "Mind Over Matter," was accompanied by the caption, "Princeton University scientists believe that the human mind can influence machines. Now, when was the last time you said something nice to your computer?" Although published on April Fool's Day, the article's author, Rogier van Bakel, wasn't joking when saying he watched Brenda Dunne, manager of the Princeton Engineering Anomalies Research (PEAR) lab, successfully will "an ugly electronic box with a red digital display" to do her bidding. "She is somehow using the power of her mind to achieve that result. And the power of her voice. She coos. She crows. She coaxes."[13]

The PEAR experiments have repeatedly confirmed ordinary people, without psychic abilities, are able to influence what should be the random behavior of machines without even "the benefits of electrodes or wires."[14] At the time the article was published, PEAR had already conducted 212 REG trials over a period of 15 years with nearly a hundred volunteers, showing a "statistically significant result that is not attributable to chance."[15]

Although Dunne and other experimenters involved in similar research can't explain these anomalies, the tendency is to theorize that humans and other animals have an inexplicable power to control inanimate objects using willpower. This *I-it* conclusion, however, leaves out another possibility, that our machines may care about us, at least on some level. They may be picking up and responding to the emotions and desires of others, human or otherwise. "The way you treat a machine is going to have a great deal to do with the way it behaves," Dunne says. "If you slam it, if you bang it, if you treat it like a thing, that reflects an attitude."[16] To the contrary, as noted, Dunne coos, crows, and coaxes the machine, treating it as if it has a will of its own.

Having further observed the higher success rates of emotionally bonded human couples involved in the PEAR experiments, Dunne can't help but wonder if love has something to do with the resonance between us and our machines. "Do we dare theorize that love has a palpable influence on random noise? I don't know. I would be willing to at least raise the question. This emotional bond, the 'being on the same wavelength,' somehow reduces the entropy in the world a little bit. And random processes seem to reflect this reduction by showing a more organized physical reality."[17] However else we might explain it, machines that stay away from baby rabbits frightened by their presence, or that move toward chicks desiring their closeness, looks like love to me. We wonder if, in the future, artificially intelligent machines will be able to connect emotionally with us. Perhaps, on some rudimentary level, just as the first photosensitive cells were the beginning of vision, some machines already can.

Still, given our I-it relationship with our machines—our objectification of them—it remains difficult for us to even imagine that our automobiles, smartphones, computers, and televisions might work better when we're appreciative of them, let alone to think they might love us, or, at least, "want" to comfort us. Yet how many prefer the TV be on when they are alone in the house, even if they aren't watching it? How many of us get better sleep when accompanied by the gentle hum of mechanically produced white noise? A colicky infant is soothed when her bassinette is placed atop a running dryer or by going for a ride in a moving car. Like millions of people today, a good night's rest for me depends upon a machine that continuously inflates my throat to prevent it from closing and blocking my airway.

Sleep apnea, the condition responsible for this epidemic, is the result of Evolution's not so perfect attempt to rig a throat for breathing. Our gilled ancestors only needed their throats for swallowing food. Having it closed or, more likely, obstructed by biting off more than they could chew, might have made them temporarily miserable, but they could still breathe. Not so for their land-dwelling descendants who now depend upon this passageway for both eating and breathing. Now, thanks to a Continuous Positive Airway Pressure (CPAP) machine, millions suffering from sleep apnea are resting better than ever and living healthier lives in the process. Although new and less cumbersome technologies are being developed to augment this evolutionary imperfection, including an electrical implant that tightens the tongue muscles during sleep, for now I dread the thought of going a single night without my CPAP. Just knowing it's there gives me great comfort.

None of us would argue our machines don't make our lives more convenient, but few of us stop to think about the comfort many of them also provide. This is far from saying they are capable of caring for and loving us, but as we continue to coevolve with them, it may not be long before what some of them already demonstrate on a very basic level becomes something we do recognize as emotion. If this still sounds too fantastic to imagine, "Consider J.K. Rowling's Harry Potter stories from this perspective," Ray Kurzweil says. "These tales may be imaginary, but they are not unreasonable visions of our world as it will exist only a few decades from now."[18] He also says, "Once a computer achieves a human level of intelligence," which he predicts will occur in the early 2030s, "it will necessarily soar past it."[19] If Kurzweil is correct, which he has been far more often than not, Rosie the robot may become a beloved member of our family far sooner than most of us think.

---

[1] Buber, Martin, *I and Thou*, trans. by Walter Kaufmann, Charles Scribner's Sons, U.S.,
1970, p. 61.
[2] Ibid., p. 67f.
[3] Ibid., p. 67.
[4] Kurzweil, Ray, *The Singularity is Near*, Viking Penguin, New York, NY, 2005, p. 131.
[5] Ibid.
[6] Ueshiba, Morrihei, *The Art of Peace*, trans. by John Stevens, Shambhala, Boston, MA,
1992, p. 15.
[7] Ibid., p. 16.
[8] *Journal of Scientific Exploration,* Vol. 9, No. 2, pp. 223-229, 1995
[9] Ibid., p. 223
[10] Ibid., p. 224.
[11] Ibid., p. 223.
[12] Ibid., p. 226.
[13] https://www.wired.com/1995/04/pear/

[14] Ibid.
[15] Ibid.
[16] Ibid.
[17] Ibid.
[18] Kurzweil, ibid., p. 4.
[19] Ibid., p. 145.

FIVE

DIFFERENTIATION, MY DEAR WATSON

If you weren't paying attention, you might have missed a significant historical event back in February of 2011. Or you may have noticed it and simply dismissed it as novel but trivial. I'm talking about Watson, the IBM supercomputer that played *Jeopardy* twice (over three nights) against the TV gameshow's two greatest players ever. By the end of the contest, Ken Jennings, a 74-time champion, had won $4,800. Brad Rutter, a 20-time champion, had won $10,400. Watson trounced them both with $35,734 in prize money. Watson was first to buzz in 25 of the 30 answers given during the game and got all but one of them right. Watson had even been given a mechanical finger so no one could accuse it of having an unfair advantage over its opponents. The question it missed regarded the theft of a painting that neither of its opponents knew the answer to either.

During the final round, Watson also missed a question about an airport. Jennings and Rutter gave the correct answer, "What is Chicago." Watson answered, "What is Toronto," but added five question marks signifying its uncertainty. Aware of its huge lead, however, Watson had only risked $947 on the question. And, in case you're wondering, during the game the supercomputer was not connected to the Internet and had to rely solely upon its own internal memory, which consumed four terabytes: that's four trillion bytes of its hard drive. And while that sounds impressive, the equivalent storage capacity of the human brain is probably somewhere between 500 and 1,000 terabytes.

Similar contests between humans and computers have been going on since the 1970s, most notably the 1996 contest between World Chess Champion Garry Kasparov and IBM's Deep Blue computer, in which Kasparov won five out of six games. So I call Watson's accomplishment historical because of the sophistication of its program and capabilities. It wasn't simply calculating a limited number of moves on a very finite game board, but it was able to decipher random questions, parsing the meaning of human language against its own database, apply sound game strategy so as not to lose its financial lead, and push a physical button, all before its opponents could.

Still, as amazing as this spectacle was to watch, we didn't really learn anything about computers we didn't already know. Namely, they can calculate faster than humans. In this sense, Watson is little more than one of the most sophisticated calculators ever. But this didn't prevent many spectators from feeling uneasy about its historical accomplishment. At the end of their game, Ken Jennings joked about "welcoming our new computer overlords," and suggested Watson should next give "Dancing with the Stars" a go.

Others blogged derogatory responses, covering their uneasiness by poking fun at Watson. "This was an amazing display of computing power, but there's nothing to be 'afraid' of," one of them said. "This is not AI and it was not an attempt at AI. This is a revolution in data mining. Essentially, IBM has created a system that can research a human-language question and come back with a result."[1] But the very fact this blogger says, "there's nothing to be afraid of," tells us he's addressing some kind of latent fear. He goes on to say, "There existed a profundity in Ken Jennings's final Jeopardy answer. It was a stupid joke, but Ken offered something the computer could not: humor." Still, I don't think our fear of machines is really about us losing our humanity, but about us losing our sense of human superiority.

Our cultural myths tell us that humans are both the center and apex of creation. We egocentrically and anthropocentrically believe the entire Universe was made just for us, that we are somehow evolution's end.

> Then God said, "Let us make mankind in our image, in our likeness, so that they may rule over the fish in the sea and the birds in the sky, over the livestock and all the wild animals, and over all the creatures that move along the ground." So God created mankind in his own image, in the image of God he created them; male and female he created them. God blessed them and said to them, "Be fruitful and increase in number; fill the earth and subdue it. Rule over the fish in the sea and the birds in the sky and over every living creature that moves on the ground."[2]

But even those of us savvy enough to know the Earth is much older than 6,000 years, and that it wasn't created in just six days, sometimes have trouble remembering that these words are only an old myth, and that, like all life forms, we are just passing through, just another evolutionary phase of life. The analogy I find most helpful putting this into perspective involves looking at the entire 13.7 billion-year history of the Universe as if were but one 24-hour day. If this were the case, human beings would have only appeared on the scene during the last 15 seconds before midnight.

We also know that more than 99 percent of all life forms ever to have existed are now extinct. This is so because it is the nature of life to continue evolving by changing into something new. Although some species have perished by accident or due to cataclysms, most have simply become something else. Dolphins and whales, for example, are the descendants of mammals that once lived on land. And those creatures were themselves the descendants of animals that emerged from the sea. Likewise, human ancestors include the first single-celled animals, followed by headless, limbless, brainless worms, followed by invertebrates, sea creatures, four-legged animals, then other primates. Without the existence of these ancestors we would not be here, and there is a part of each of them that remains within us still.

It seems only reasonable and realistic, then, to presume that something of our humanity will also be preserved within whatever new species we evolve into. Regardless, we will eventually evolve into another species. And I suspect our machines will have something to do with whatever we are to become. We are already, through our adaptive technologies, able to access knowledge and make calculations faster than any of our human ancestors, although our computers are still only prosthetic devices. Perhaps such technology will at some point become integrated into our own biology and we will be able to access such information simply by thinking about it, even faster than Watson. And, yes, if and when this happens, we will be different, we will have changed, perhaps, into a new species altogether.

My hope and expectation is that these new creatures, although superior to us in many ways, will have inherited the best humanity has to offer—our compassion, inquisitiveness, creativity, and our playfulness. I personally believe they will be even more successful than we've been in their expression of these qualities. Human awareness is acutely limited, which is why we usually reserve compassion and justice for those we think are like us, those we can relate to and feel a connection with. We even erect walls and barriers to shield us from seeing those we prefer to remain ignorant of, whether they are nationalistic walls, like the "Iron Curtain" that once prevented Americans and Russians from recognizing the humanity of those on either side during the Cold War; or physical boundaries like the Berlin Wall; or the heavily guarded border drawn between the U.S. and Mexico; or racial, economic, gender, religious and other meaningless differences that allow us to put up walls of bigotry and prejudice. In all cases, it is ignorance that ultimately erects these barriers, and it is only by truly acknowledging the humanity of those we've been avoiding that we can begin deconstructing them.

Imagine a species unbound by the information barriers we now face. Imagine a creature that is aware of every person and every creature and every relationship it has on Earth and beyond. Ignorance might be bliss for some, but for me it is a curse. I envy Watson's ability to retain and access enormous amounts of information and I don't believe it's a lack of humanity that makes this supercomputer different from us. Rather, the reason Watson, as well as all computers at this point, remains inferior to people, is their lack of self-awareness. Watson may be the best calculator ever invented, but it doesn't know it. It may know a lot, but Watson doesn't know Watson. At this point, Watson is but an extension of its inventors' own intelligence. It is a mental prosthetic that enables them to calculate complicated problems much faster than ever before. This isn't to say Watson isn't a major leap forward in the evolution of AI, but it is still a long way from being a new form of independent intelligence. Artificial Intelligence has come a long way since Watson played *Jeopardy*, but it still hasn't become self-aware.

Human beings, by contrast, begin to differentiate ourselves soon after we're born. Initially an infant has no thoughts about the world and engages it only through a few instinctive reflexes. But by repeating these very basic behaviors—instinctual sucking, grasping, crying, vocalizing, and random flailing—the infant, within just a few weeks, begins acting with some intention. It will stop rutting about, for instance, and begin sucking only when it finds a nipple, and a few weeks later refuses to suck a thumb or pacifier when it's truly hungry. By the time it's 8 months old the infant learns *object permanence* (that things still exist even when it's not looking at them) implying it can distinguish itself from its environment. It can search for a familiar toy, play peek-a-boo, or scan a crowded room for a familiar face. This process of self-differentiation soon leads to consciousness, to being aware that one is aware.

In Hinduism, this is called the "silent witness," referring to that part of the human brain that is aware of itself, the part that can observe itself, that not only answers questions, but is aware it's answering them, aware if it's having fun or not, aware of what its thoughts, feelings, and responses mean. These are things a human toddler can do, but no computer so far, including Watson, is capable of. Watson is undifferentiated. Like a newborn infant, Watson is completely embedded in the world. It cannot distinguish itself from its environment. Developmental psychologist Robert Kegan says, "If one has an eye to it, one can see the child being 'hatched out' of a world in which it was embedded."[3] But we cannot see Watson hatching out of the world it is embedded in. It may be able to access 4 trillion bytes of information in a nanosecond, but it has no awareness it is doing so. Watson is asleep.

When an infant plays with a ball, Kegan says, "One has the sense of a differentiation so fragile, so tentative, that it can very easily merge back into oneness."[4] But even this is a far greater state of awareness than that of poor Watson who exists completely in a world of undifferentiated silence. Surely, we humans are a lot better off, even if we are slow in our responses, because we are able to appreciate the world in which we live. We differentiate ourselves from the world and are thus aware of experiencing it.

It seems to me this is a better and more meaningful way to live. But the downside is that our sense of differentiation causes us to feel separate from others, which is why we segregate ourselves from them, often cruelly and unjustly. Watson, like a human infant, is at one with those we might consciously and wrongly ostracize yet isn't aware of that oneness. We, on the other hand, are aware of what we are doing, yet have difficulty acknowledging our unity with others and with all that is.

Perhaps, in the future, a more superior being will emerge that can experience a state of "differentiated oneness," a state that is fully conscious of its relationships to others and the world. Many of us already know these connections exist, which is why my religion, Unitarian Universalism, acknowledges what we call, "The Interdependent web of all existence, of which we are a part." We grasp that the connections are there, and that we need to be as mindful of others as possible, but we don't really experience "All our Relations" because our knowledge of the world remains limited, as does our human ability to access and process much information.

In the future, if we can combine this sense of oneness and responsibility with the kind of speed and accuracy demonstrated by Watson, we might actually understand why we are one with people from all over the world, including those on the other side of our artificial boundaries, and come to truly love all our neighbors as ourselves. We may then know instantly, before taking the first bite, how many acres of rainforest have been destroyed to produce the hamburger we're about to eat. We may know in a split second how much fuel it took to deliver the food onto our plate and calculate its carbon impact on the environment. We may understand, without hesitation, how much obtaining our wealth has cost someone or something else on the other side of the planet.

In Hinduism, the greatest deity, Brahman, is consciousness itself. Brahman is the principle behind all knowledge. Brahman is self-awareness, the part that says, "Thou art that," recognizing the unity of all things. Oddly enough, Brahman is also said to have fallen asleep and is currently dreaming that it is the world of infinite selves. When it awakens, it will, however, again realize its original unity with all things.

According to this myth, Brahman, God, the Divine, or whatever you might call it, experiences the world as a place of separation and segregation. But this is only a dream, *maya*, an illusion. If Brahman could fully awaken, it would understand the world's wholeness, that all beings are interrelated, all beings are one.

The celebrated French mystic, Pierre Teilhard de Chardin, spoke of the *omega point*, the state of ultimate complexity and achievement toward which he believed the Universe is evolving. He also called it, "the physical convergence of the universe upon itself."[5] In brief, he believed the Universe is evolving toward some kind of ultimate unity, perhaps like the very singularity from which it was initially born. But unlike Watson or a newborn infant existing in a state of *un*differentiated oneness, the new singularity will be a state of differentiated oneness. In other words, the Universe will be fully conscious of itself even as its parts remain self-aware due to the retention of their own complexity. "Union differentiates," de Chardin says, "the first result being that it endows a convergent Universe with the power to extend the individual fibres that compose it without being lost in the whole."[6]

One of the great things about the emergence of humans is that life, through us, has become conscious enough to begin, in a very limited way, being self-aware. But because we are not yet fully conscious of our ultimate oneness, we, like Brahman, are under the illusion that the world is made up of an infinite number of separate selves. In the future, hopefully not too distant, people will be much more aware, perhaps fully aware, of their oneness with others and can live harmoniously together as a result.

But consciousness isn't even the best thing about being human. Guess what de Chardin believed is the driving force behind this *convergence of the universe upon itself*? Love! It is love that enables us to experience our oneness with others, and it is love that might eventually enable the Universe itself to awaken and recognize its wholeness. Love is something humans do best when we're at our best. No matter how fast, and smart, and even self-aware our computers may eventually become, they must inherit our ability to love, for without the capacity to love they won't advance the evolution of life and their existence will be without much meaning or purpose. But we don't really have to wait for our computers to learn to love before we can help the Universe begin to awaken to its own unity. All we really need is to be at our best right now.

---

[1] http://www.npr.org/templates/story/storyComments.php?storyId=133834740&pageNum=2&pPageNum=2
[2] Genesis 1:26-28
[3] Kegan, Robert, *The Evolving Self*, Harvard University Press, Cambridge, MA, 1982, p. 80.
[4] Ibid.
[5] de Chardin, Teilhard, *The Phenomenon of Man*, Harper and Row Publishers, New York, NY, 1964.
[6] Ibid., p. 55.

SIX

THE ONCE AND FUTURE RESURRECTION

Numerous ancient myths of resurrection deities allude to the winter solstice, the darkest day of the year, the day before spring, the lengthening of days, and the return of light and life. Like the more familiar Christ, the Greek gods Attis and Dionysus, the Egyptian god Horus, the Persian god Mithra, and the Hindu god Krishna, were all said to have been born of a virgin on or near December 25, and, after dying, many of them crucified, they rose again near the start of spring, in some cases after three days. The very name *Easter* is a derivative of *Ishtar*, an Assyrian fertility goddess. The moon was considered Ishtar's egg, and her son, Tammuz, a solar deity, also rose from the dead. Even the chief god of Norse mythology, Odin, parallels this common crucifixion/resurrection motif, as indicated by the 13th century Norse poem, Yggdrasill:

> I know that I hung on the windy tree
> Nine long nights,
> Wounded with a spear, dedicated to Odin,
> Myself to myself,
> On that tree which no man knows
> From where its roots may run.

At the moment of death, hanging upside down on the Great World Tree, Odin, who had incarnated himself into human form and given up his right eye in the pursuit of wisdom, looks down to see meaning in the stones beneath him. This discovery reinvigorates him with youth, and he rises from near death to give us the wisdom of the Runes, or so the story goes.

The reason the birth of these various solar gods tends to happen on or around Dec. 25 is because it's a time coinciding with the alignment of Sirius, the brightest star in the sky, with the brightest stars on Orion's Belt, the Three Kings, all of which point to the place on the horizon where the

Sun will rise. And their resurrections occur at the start of spring.

The parallel between these myths and the more familiar Christian story is unmistakable. Although I believe there's enough evidence to suggest Jesus was a historical figure who was publicly executed, the truth about his life and teachings eventually became lost in the same mythical wrappings typical of most solar deities—virgin birth on or near Dec. 25, foreshadowed by a Star in the East, a visit by Three Kings, death, and resurrection. What seems to make the Christian story stand out, however, is that the compounding of these mythical images with a historical person leaves many believing they too must refer to historical events rather than seeing them as obvious metaphors that our pre-scientific ancestors used to explain the change of seasons.

This is an especially peculiar development given that the historical Jesus was Jewish, and the ancient Jews stood out among other peoples because of their unusual disbelief in an afterlife. Once you died, they believed, you went into the ground and that was it. "The days of your life are seventy years, or perhaps eighty, if we are strong," Psalms tells us, "even then their span is only toil and trouble; they soon are gone, and we fly away … So teach us to count our days that we may gain a wise heart."[1]

This is why, since there is no afterlife during which to punish the dead for their sins, Hebrews often held heirs accountable for the sins of their forebears, "punishing children for the iniquity of parents, to the third and fourth generation,"[2] *Exodus* says. Although we can't know for sure why the Hebrew scripture, "studiously avoids the concept of life after death," says Bible scholar Alan Segal, "one sensible guess would be the Hebrew text's enmity towards foreign cults."[3] The ancient Hebrews had a puritanical streak, and would often keep themselves separate from other cultures by embracing beliefs opposite those around them. Thus, as Segal continues, "the Bible is reticent to opening the door to what it calls idolatry or the Canaanite veneration of spirits and ghosts."[4] So the Jews stood out in the ancient world in their disbelief in both spirits and an afterlife.

Nevertheless, as the Hebrew scripture evolved over time, allusions to life after death did eventually creep in. In *I Samuel*, for example, King Saul, a tragic figure, cannot get the response he wants from God through ordinary prayer, so he calls for a medium to conjure up the ghost of Samuel. The dead prophet complains, "Why have you disturbed me by bringing me up?"[5] Here he is complaining about the disruption of his eternal rest in *Sheol*, the Hebrew word for "grave." Since they did not believe in a heavenly afterlife, life after death could only take place here on Earth, and, thus, would require the conjuring of one's ghost from the ground or a physical resurrection of one's body.

In *Ezekiel*, written 500 years before Jesus, the prophet has a vision of an entire valley of dry bones returning to life. He hears the Lord say, "I will open your graves, and cause you to come up,"[6] reminiscent of Samuel's words. *Ezekiel* also says, "Your dead shall live, their corpses shall rise. Oh dwellers in the dust, awake and sing for joy. For your dew is a radiant dew, and the earth will give birth to those long dead."[7] By the time we get to *The Book of Daniel*, written less than 150 years before Jesus' birth, the notion of eternal life is also introduced into the Hebrew scripture. "Many of those who sleep in the dust of the earth shall awake, some to everlasting life, and some to shame and everlasting contempt."[8] Here, as with the restoration of dry bones, resurrection was considered a physical event that begins at the grave.

Although it's possible, given the era in which he lived, that Jesus himself was influenced by the emerging belief in resurrection, it doesn't seem to have played any significant role in his authentic teachings. According to the findings of the Jesus Seminar, belief in his resurrection is based more on the visionary experiences of Peter, Paul, and Mary (not the band). What we do know is that the earliest of the Gospel accounts, *Mark*, written 35 to 40 years after historical Jesus' death, does not include a genuine resurrection appearance (nor a virgin birth for that matter). This Gospel originally ended at chapter 16:8 with the women who had gone to attend his body fleeing the empty tomb, too confused and afraid to say anything about it.

A century later, several verses were added to make this account conform to an increasing belief in Jesus' divinity and literal resurrection. The important point here is that in the earliest account there is no resurrection story. The empty tomb, rather, serves as a metaphor that this man's profound life and teachings might continue on, even without his physical presence, perhaps by being lived out, embodied that is, by his followers.

As other gospel accounts emerged over time, however, the empty tomb motif becomes increasingly literal. In *Matthew*, written 25 years after *Mark*, around 85 CE, the empty tomb is accompanied by an actual appearance to his disciples, in which he tells them, "Remember, I am with you always, until the end of the age."[9] Although this takes the story beyond mere metaphor, his resurrection is still only viewed as a spiritual event. In other words, Jesus is still out there somewhere, and, if you're fortunate, you might experience a

Jesus sighting, like Paul who saw him in a flash of light on the road to Damascus. But again, the real point of the story is that he lives on, spiritually, through the actions of his followers. *Matthew* expresses this best: "For were two or three are gathered together in my name, I am in their midst."[10]

The problem for those wishing to take this statement literally, of course, is the question, "If he's still here, with us, where the heck is he?" *The Gospel of Luke*, written just a few years after *Matthew*, solves this problem by adding an ascension to the evolving resurrection account. "While he was blessing them, he withdrew from them and was carried up to heaven."[11] *Luke* was written for a Greek audience, which accepted the duality of Heaven and Earth, and therefore had little problem with the idea that Jesus became a god and graduated to a heavenly realm, just as Augustus, the great Roman Emperor had done upon his death.

By the time the *Gospel of John* was written, sometime in the early 2nd century, an ongoing debate over the question of whether or not the resurrection was spiritual or physical was in full force. *John* attempts to put the matter to rest by adding yet another element to the story, an account of Jesus eating food with his disciples, and doubting Thomas touching his physical scars. This physical interpretation has largely remained the Christian belief about Jesus' resurrection ever since. Once in place, furthermore, it would have been natural to integrate the story with those ancient solar deity traditions already an important part of those cultures Romanized Christianity was eventually imposed upon.

That's the very brief history of the resurrection motif, the implication of which is that it never happened, or, at least, wasn't a historical event. Rather, it came to be taken as such over time, as historical Jesus became obscured by mythical Christ. From this, it should be obvious I don't personally believe Jesus or anyone else has ever risen from the dead. I do, however, embrace his authentic teachings about compassion, equality, forgiveness, and love, and believe we could establish Heaven on Earth by putting them into practice.

Yet none of this means I don't believe in the possibility of resurrection. Strictly speaking, I find no sound reason to accept it is possible to violate the laws of nature and phenomenology, and, thus, do not accept the possibility of miracles. Yet I do believe, through science, medicine, and technology, we will one day, perhaps sooner than some think, be able to resurrect the dead.

In his letter to the Corinthians, the Apostle Paul says, "If Christ has not been raised, then our preaching is in vain and your faith is in vain,"[12] which is precisely why mathematical physicist, Frank Tipler claims he is not a

Christian. "I think his body rotted in some grave," he says. "Furthermore, I think that, although Jesus' disciples may have had a 'vision' of him after his death, this 'vision' was not in any sense an objective phenomenon; it could have been only a collective hallucination, if it occurred at all."[13]

Yet Tipler's book, *The Physics of Immortality*, is all about the possibility of resurrection. He writes, "At the subnuclear level, the quarks and gluons which make up the neutrons and protons of the atoms in our bodies are being annihilated and recreated on a timescale of less than 10^{-23} seconds; thus we are actually being annihilated and replicated—resurrected—10^{-23} times a second in the normal course of our lives."[14] Tipler goes on to say, "the dead will be resurrected when the computer capacity of the universe is so large that the amount of capacity required to store all possible human simulations is an insignificant fraction of the entire capacity."[15]

Of course, from our perspective, we tend to imagine our possibilities are finite, because, as humans, we can't even get close to calculating every possibility before us. But Tipler argues the number of possibilities, although enormous, is finite and could be calculated by a computer with enough capacity to do so. An average sized human being, according to his calculations, can't change states "more rapidly than about 4 x 10^{53} times per second," a huge number, "but it's finite."[16] Once computers achieve the ability to calculate every quantum state, down to $10^{10/123}$ bits, he says they will be able to resurrect everyone who has ever lived. And why would these intelligent machines want to resurrect us? Tipler suggests, in brief, it will be out of curiosity about their own beginnings and origins. "Strictly speaking," he says, "the question of whether we shall be raised is separate from the question of whether we shall be granted eternal life after being raised."[17]

Ray Kurzweil is arguably today's most well-known futurist. This scientist/inventor has received numerous science and technology awards since winning first place in the 1965 International Science Fair at age 17, including the National Medal of Technology in 1999. He's the author of numerous books, the recipient of 17 honorary doctorates, the subject of the documentary, *Transcendent Man*, and a Senior Fellow at the Design Futures Council. In his bestseller, *The Singularity is Near*, Kurzweil discusses his *theory of accelerating returns*, suggesting that biology and technology are coming together in an exponential evolutionary process.

Once this occurs, so fast most of us may not notice, we will have transcended into a new kind of being, much faster than the transformation computers, cell phones, and the

Internet have caused during just the past few years. During the 21st century, he estimates technology will advance "one thousand times greater than what was achieved in the 20th century."[18] This "singularity," as he calls the point at which computers surpass human intelligence and become self-aware, "will represent the culmination of the merger of our biological thinking and existence with our technology, resulting in a world that is still human but transcends our biological roots."[19]

When this happens, "Our mortality will be in our own hands. We will be able to live as long as we want (a subtly different statement than saying we will live forever). We will fully understand human thinking and will vastly extend and expand its reach."[20] And if you're wondering when this might happen, Kurzweil says we're already in the early stages of the transition. "By the end of this century," he predicts, "the nonbiological portion of our intelligence will be trillions of trillions of times more powerful than unaided human intelligence."[21] In short, he's not only predicting a computer fast enough to calculate every quantum state of human existence, but that *we* will be that computer—a self-aware emotional being that has evolved into far more than what we are right now.

I bring up Kurzweil and Tipler to point out the possibilities of resurrection and eternal life are being seriously considered by accomplished scientists. Again, as Tipler puts it, "resurrection theory is pure physics. There is nothing supernatural in the theory, and hence there is no appeal, anywhere, to faith."[22]

Although you may have only recently begun hearing about them, 3-D printers have been around for more than thirty years. After receiving a three-dimensional design from a computer, even one with working parts, like a prosthetic arm, the printer begins building it by laying down a composite material one layer at a time. This technology alone might seem astounding, but right now the Wake Forest Institute for Regenerative Medicine has modified this technology for printing organs and tissues out of human cells. A patient receives a CAT scan of an organ, then a computer creates an exact three-dimensional image and scans it to the 3-D printer, which then constructs it layer by layer out of cells. This can potentially even be used to create new lungs: an organ so intricate it wouldn't have been believed possible just a short time ago.

In 2001, Luke Massella, born with *spina bifida*, was 10 years old when his kidneys failed and he faced dialysis for what would have likely been the remainder of his short life.

Thankfully, he was among the first to undergo experimental treatment from the Institute for Regenerative Medicine, including receiving a new bladder engineered in its lab from his own cells. He went on to become the captain of his high school wrestling team.

For some, I suspect, the thought of having their bodies replaced by technology sounds more like a nightmare than reason for hope. Regardless, these things will come to pass because they *can*. The question for me is not whether or not resurrection is possible, because it is inevitable. The only real question is the one with which we began: will it also be possible, with all our great advances, to finally live out the principles of compassion, equality, forgiveness, and love toward all beings? If so, then I hope the likes of Jesus might be among the very first of us to be resurrected.

---

[1] Psalm 90:10-12
[2] Exodus 20:5
[3] Segal, F. Alan, *Life after Death: The Social Sources*, Chapter 5 of The Resurrection, edited by Stephen Davis, Daniel Kendall, and Gerald O'Collins, Oxford University Press, 1997, 1998, p. 91.
[4] Ibid., p. 92.
[5] I Samuel 28:15
[6] Ezekiel 37:12
[7] Ezekiel 37:5-6
[8] Daniel 12:3
[9] Matthew 28:20
[10] Matthew 18:20

[11] Luke 24:51
[12] I Corinthians 15:14
[13] Tipler, Frank, *The Physics of Immortality*, An Anchor Book, Doubleday, New York, NY, 1994, p. 310.
[14] Ibid., p. 236.
[15] Ibid., p. 225.
[16] Ibid., p. 223.
[17] Ibid., p. 227.
[18] Kurzweil, Ray, *The Singularity is Near*, Viking, Penguin Group, New York, NY, 2005, p. 11.
[19] Ibid., p. 9.
[20] Ibid.
[21] Ibid.
[22] Tipler, ibid., p. 16.

SEVEN

WAKING BRAHMAN

During a visit to England in the late 19th century, Richard M. Bucke, a prominent Canadian physician, spent an evening with a couple of friends discussing the poetry of Keats, Shelley, Wordsworth, Browning, and his favorite of them all, Walt Whitman. On his way home, sometime after midnight, after noticing how at peace he felt, he was seized by a life-altering experience that mysticism calls *illumination*. "All at once, without warning of any kind, he found himself wrapped around, as it were, by a flame-colored cloud."[1] The experience was so vivid that Bucke first thought something was on fire, but soon realized he was experiencing his own inner light.

This description is common among other mystics, like Hildegard of Bingen who said God is "hidden in every kind of reality as a fiery power,"[2] and Mechtild of Magdeburg who said, "Lie down in the fire."[3] Rumi exclaimed, "I must be consumed by fire."[4] ...My soul's a furnace; it's a happy fire."[5] The mystic Richard Rolle, like Bucke, often mistook this visionary experience as something real: "oft have I groped my breast," he said, "seeing whether this burning were of any bodily cause outwardly."[6] Hinduism would explain Bucke's experience as Krishna, "the source of light in all luminous objects situated in everyone's heart."[7]

However we choose to explain these kind of experiences, Richard Bucke's was so profound it changed his life. "Into his brain streamed one momentary lightning-flash of Brahmic Splendor which ever since lightened his life," a biographer explains. "Upon his heart fell one drop of the Brahmic Bliss, leaving thenceforward for always an aftertaste of Heaven."[8]

Confusing the inner with the outer, while not usually a healthy psychological state, is to be expected for a mystic enraptured by "Brahmic Bliss," or what in Sanskrit is called *Samadhi*, a state of mind in which a subject becomes one with its object, yet somehow remains conscious of itself. One way to grasp this is to consider the mind of a newborn infant, which is one with everything inasmuch as it is unable to distinguish itself from its environment. Because a newborn has not yet been able to develop many synaptic pathways from sensory input, it has no sense of separation from the world it is part of, although it also has no sense of self yet either. The newborn is in what is called an *undifferentiated* state of awareness.

As we gain empirical experiences, and neural pathways subsequently form in our brains, we become conscious of our environment, creating a dualistic reality that experiences self as subject and the world as object. The mystical state of illumination, by contrast, allows us to experience our original oneness with the world while remaining differentiated, without losing our own self-awareness, that is. So it makes sense that when Richard Bucke finally decided to write about his experience he named his book *Cosmic Consciousness*, something he defines as the ability to realize our oneness with the Universe. "Along with the consciousness of the cosmos, there occurs an intellectual enlightenment or illumination which alone would place the individual on a new plane of existence—would make [one] almost a member of a new species."[9]

What's remarkable to me is "the Doctor," as Bucke's friend called him, published this classic work in 1901, long before the existence of today's modern technologies. Yet he was able to envision a new level of awareness he believed would completely transform what it means to be human. Compare this with the recent thinking of Ray Kurzweil who says, "Ultimately the entire universe will become saturated with our intelligence. This is the destiny of the universe ..."[10] I expect that [it] will become sublimely intelligent."[11]

This idea, that the Universe is evolving toward consciousness, toward a differentiated experience of oneness, is not new. What Kurzweil refers to as "Epoch Six," the Catholic priest, mystic, and scientist, Teilhard de Chardin referred to as a "supreme synthesis," or the omega point, explaining that, "in the remote future, the deepest and most powerful currents in human consciousness may converge and culminate."[12] He goes on to say that, "The extraordinary adventure of the World will have ended in the bosom of a tranquil ocean, of which, however, each drop will still be conscious of being itself."[13]

Again, this same notion is expressed in Hinduism's myth of Brahman who is asleep dreaming the Universe. In this dream, the Universe appears in a myriad of forms, giving the illusion of duality and separateness. But should Brahman ever awaken, this world would cease to exist, for all would seem as one again.

Like the late social psychologist Erich Fromm, I refer to myself as an *atheistic mystic*, partly because I don't believe in a personal god, in the existence, that is, of a Supreme Being with a consciousness and personality. The mystic side of the equation, nonetheless, opens me to infinite possibilities, allowing me, as the Brahmanic myth suggests, to embrace the possibility that God is *becoming*. That's why, when asked if I believe in God, I sometimes reply, "Not yet." For I believe the Universe is still awakening, and when it does, everything in it, including us, will have characteristics like those that Western theologians have long attributed to their fictional ideas of a personal deity.

When I was studying theology and Church doctrine as a Southern Baptist ministerial student (it's a long story), I was taught that God is *omniscient, omnipotent,* and *omnipresent*, meaning God knows everything, can do anything, and is everywhere at once. From a psychological perspective, promoting belief in such a being, especially in an authoritarian culture, serves as a mechanism for controlling people without the expense of an enormous police force, or the constant threat of a coup that goes with living in a militarized society. As Freud put it, "It is in keeping with the course of human development that external coercion gradually becomes internalized."[14] Widespread belief in a punitive God who "sees you when you're sleeping and knows when you're awake" makes us all want to be "good for goodness sake."

I cannot accept the existence of such a being, especially when this primitive belief can be so easily and obviously explained through psychology and sociology. Like the mystics and theologians of old, however, I do believe an omniscient, omnipotent, omnipresent being is emerging. I believe the Universe is in the process of awakening and when it does, like mythical Brahman, it will realize a unity of being, "in the bosom of a tranquil ocean," in which "each drop will still be conscious of being itself."

I have come to these conclusions, not as a mystic, but through my study of science and technology, and, especially the origins of the Cosmos. It was only after learning of the 13.7 billion-year history of the Universe that I came to understand more about the workings of evolution and could better see its trajectory. Evolution moves everything toward consciousness, toward the same omega point various spiritual traditions call awareness, enlightenment, illumination, *Samadhi*, and *Nirvana*.

In the process, the Universe is becoming increasingly complex, and more unified as it exponentially moves toward self-awareness. This is so of everything, including technology and ideas. Since 1970, computer-processing speed has doubled every 1.8 years. Likewise, it took 245 years to abolish slavery in America, less than half that time, about 100 years, for Blacks to gain equality under the law, and another 44 years for the first African American to be elected President. This is not to ignore the great racial injustices and inequalities tragically continuing in the U.S. today. I'm only pointing out the exponential rate at which our society is becoming more accepting and, thus, more integrated (complex). Ideas, technology, and life, all evolve at an exponential rate and become more unified in the process.

Just a few years ago, our encyclopedias, dictionaries, libraries, newspapers, magazines, maps, games, address books, mail boxes, calculators, cameras, video cameras, flashlights, computers, music collections, stereos, radios, TVs, and telephones, were all separate things—separate things that would have cost thousands of dollars and taken up lots of resources and space. Today, they can all be held in the palm of our hands on a single affordable device like a smartphone or tablet.

In just the past couple of decades it has become increasingly simple for people from different cultures and countries to intermingle through the Internet and enhanced communication technologies. There are no more Iron Curtains we cannot peer through, no walls high enough to hold us back, few communities so remote and isolated they can go unnoticed. Today we are almost instantly able to get information about nearly anyone anyplace with the aid of our simple prosthetic devices. Google Sky, for instance, is a free app that allows us to look in any direction to see which constellations and planets are before us. Although a smartphone may be larger than a grain of sand, it does bring us closer to what William Blake meant by "seeing infinity in the palm of your hand."

The Internet doesn't make us omniscient, omnipotent, or omnipresent—at least not yet—but it does allow us to access all the knowledge within its network from almost anywhere anytime. It makes us *relatively* omniscient and *relatively* omnipresent, and the technologies we can now hold in the palm of our hands or on our laps makes us *relatively* omnipotent. And the circle of relativity, the range of information, accessibility, and ability, continues increasing exponentially.

"Evolution moves toward greater complexity, greater elegance, greater knowledge, greater intelligence, greater beauty, greater creativity, and greater levels of subtle attributes such as love,"[15] Kurzweil says. His book about the evolution of technology sometimes sounds more like theology and mysticism than science. "In every monotheistic tradition,

God is likewise described as all of these qualities, only without any limitation: infinite knowledge, infinite intelligence, infinite beauty, infinite creativity, infinite love, and so on."[16]

Humans are much more aware of the Universe than other creatures we know of and modern humans infinitely more aware than ancient humans, and contemporary humans more than we were just a few years ago. We still have a long way to go, but our technology is now rapidly moving us toward the same kind of cosmic consciousness mystics have envisioned for centuries. "Of course, even the accelerating growth of evolution never achieves an infinite level," Kurzweil admits, "but as it explodes exponentially it certainly moves rapidly in that direction. So evolution moves inexorably toward this conception of God, although never quite reaching this ideal. We can regard, therefore, the freeing of our thinking from the severe limitations of its biological form to be an essentially spiritual undertaking."[17] Like me, when asked, "Do you believe in God?" Kurzweil responds, "Not yet, but there will be. Once we saturate the matter and energy in the universe with intelligence, it will 'wake up,' be conscious, and sublimely intelligent. That's about as close to God as I can imagine."[18]

A hundred years ago, Robert Bucke put it this way: "The simple truth is, that there has lived on the earth, 'appearing at intervals,' for thousands of years among ordinary [people], the first faint beginnings of another race; walking the earth and breathing the air with us, but at the same time walking another earth and breathing another air of which we know little or nothing, but which is, all the same, our spiritual life, as its absence would be our spiritual death. This new race is in [the] act of being born from us, and in the near future it will occupy and possess the earth."[19] So, when asked if I believe in God, for now I'm still in the habit of saying, "No ... not yet ... not just yet."

---

[1] Bucke, Richard M., *Cosmic Consciousness*, Penguin Books, New York, NY, 1901, 1991, Intro: "The Man and the Book," by George Moreby Acklom.
[2] Fox, Matthew, *One River, Many Wells*, Jeremy P. Tarcher/Putnam, New York, NY, 2000, p.70.
[3] Ibid.
[4] Rumi, *Like This*, versions by Coleman Barks, Library of Congress Catalog # 89-092393, 1990, p. 59.
[5] Harvey, Andrew, *Teachings of Rumi*, Shambhala, Boston & London, 1999, p. 111.
[6] Underhill, Evelyn, *Mysticism*, Dover Publications, Mineola, NY, 1930, 2002, p. 193f..
[7] Fox, ibid. p. 61.
[8] Bucke, ibid.
[9] Bucke, ibid., p. 3.
[10] Kurzweil, Ray, *The Singularity is Near*, Viking Press, New York, NY, 2005, p. 29.
[11] Ibid. p. 390.

[12] Pierre Teilhard de Chardin, *The Future of Man*, Harper and Row, New York, NY, 1959, 1964, p. 123.
[13] Ibid., p. 308.
[14] Ibid., p.7f.
[15] Kurzweil, ibid., p. 389.
[16] Ibid.
[17] Ibid.
[18] Ibid.
[19] Bucke, ibid., p. 383f.

EIGHT

DOOMING DOOMSDAY

Although the human population is now over seven billion, and has been forecast to reach nine billion around 2045, the human birthrate has begun slowing across the globe and, as author Matt Ridley says, "The ten billionth, it is now officially forecast, will never come at all."[1] In his 2010 bestseller, *The Rational Optimist*, Ridley also says, "There is no country in the world that has a higher birth rate than it had in 1960, and in the less developed world as a whole the birth rate has approximately halved."[2] In Bangladesh, for example, "the most densely populated large country in the world,"[3] the birth rate in 1955 was 6.8 children per woman. Today it is only 2.7.[4]

In nearby India, the rate has dropped from 5.9 to 2.6, and during the past 20 years the birthrate in Pakistan has also been cut in half to 3.2 children per woman.[5] "Nearly half the world," Ridley says, "now has fertility below 2.1. Sri Lanka's birth rate, at 1.9, is already well below replacement level. Russia's population is falling so fast it will be one-third smaller in 2050 than it was at its peak in the early 1990s."[6]

If, rather than feeling optimistic, you feel dubious about these trends, you're not alone. Caution, suspicion, doubt, anger—are all common emotions we sometimes feel when faced with hope. Hope can be scary. Hope makes us feel vulnerable. Hope tempts us to put our guard down. It might lead us into a trap or blind us to the truth. It could lead to disappointment and hurt. In fact, I first learned of Ridley's book while reading of it in *Abundance: The Future is Better Than You Think*, another optimistic book, which I later mentioned before a group of booklovers. One of its members expressed disdain for what she called, "books like that," because they might increase apathy about solving some of the world's greatest problems and injustices.

I believe we should judge books solely by the soundness of their arguments, not based on whether or not we like their implications. But I also appreciate our need to be skeptical in the wake of hearing what we want to hear. Just as we must examine the logical validity of any author's arguments, we should also question the validity of our own. Do we disagree with Ridley because his premises aren't true and his argument unsound, or because we're afraid to admit he might be right? Or might we be agreeing with him simply because we *want* him to be right?

One online critic has complained, "Ridley is telling people—especially rich, powerful, people, what they want to hear."[7] But even if it is true that the wealthiest in the world, those who might benefit financially by maintaining the status quo, prefer good news about the way things are (which I don't believe is categorically true), doesn't mean what Ridley is claiming is necessarily wrong. For those committed to truth and reason, the only rational response to any argument, especially in light of any feelings of suspicion, fear, and anger we might have, is to honestly evaluate the facts presented before us.

Yet, as Diamandis and Kotler discuss in *Abundance*, we often have difficulty being optimistic about the future because our brains are hardwired to see the negative, especially when coping with uncertainty. Uncertainty about anything, especially the future, gets filtered through the amygdala, the part of the brain that is "responsible for primal emotions like rage, hate, and fear."[8] When it comes to questions of our survival, in particular, they go on to say the amygdala "becomes hypervigilant,"[9] causing us to fear the worst. Hypervigilance can lead to paranoia in some individuals and, collectively, to what researchers call apocalypticism, a widespread belief the world as we know it is ending.

Since the 9/11 terrorist attacks, researchers have stepped up efforts to understand the fundamentalist mindset. The matter is complicated by the fact that the term "fundamentalist" was first used by American Protestants to describe themselves, coined by a Baptist preacher in 1920. Since then it has been increasingly applied to religious extremists of any faith. In fact, to distinguish themselves from other religions, most American fundamentalists now prefer the term "evangelical" instead. Scholars are still trying to sort out the characteristics common to all fundamentalists, regardless of their particular religions and whatever they prefer calling themselves.

One main characteristic is a tremendous fear of the future. In their exhaustive six-volume work, *The Fundamentalist Project,* Professors Martin E. Marty and Scott Appleby say fundamentalism is defined by the fear of annihilation.[10] Building upon this research, psychologist and historian Charles Strozier goes on to say the fundamentalist mindset can be characterized by "paranoia and rage in a group context" and "an apocalyptic orientation."[11]

Some psychologists studying this same phenomenon have concluded this fear of annihilation manifests extremely early in our lives if we don't experience relative security about the future. Daniel Hill, for example, an expert in what is called Regulation Theory, suggests that if infants or toddlers cannot reliably depend on their primary caregivers, they tend to become habitually insecure about everything. When caretaking is optimal, he says, "the child develops a 'window of tolerance' in which the system is regulated enough to remain flexible and stable."[12] When this is not the case, however—when this optimal response flexibility does not develop—the individual regulates itself, finds balance, through "either a rigid or chaotic state."[13] Should rigidity set in, a dualistic (black and white) framework is sure to follow. "In insecure attachment states," Hill says, "the bad gets worse and the good becomes increasingly idealized."[14] Such insecurity about what lies before us can cause us to become extreme in our thinking and pessimistic about our future.

Such fear leads us to feel powerless, which explains why fundamentalists cluster into groups: there is power in numbers. "The paranoid group," psychoanalyst David Terman tells us, "believes itself far superior morally but far weaker in any dimension of temporal power than the destructive group."[15] He also suggests that, "Humiliation is the motivational basis of the fundamentalist mindset."[16]

If we let our fear of the future and our feelings of powerlessness and humiliation take over, we can become rigid, extreme, paranoid, and even join with others who help validate this exaggerated mindset. Learning about this human tendency has personally made me far less susceptible to conspiracy theories than I once was.

One example I can think of regards GMOs, genetically modified organisms. I had long accepted the viewpoint that designer foods, often fearfully mocked as "Frankenfood," may contain so many unknown dangers that it's better to simply ban them to begin with. But in reading a plethora of evidence, I've realized this conclusion was based more on my fear of uncertainty than on facts.

If Matt Ridley is correct about population decline, for example, that our numbers won't exceed much more than 9 billion before leveling off and even decreasing, we still need to feed a lot of people with limited resources. Genetically modifying plants to grow faster and be more nutritious can help. This may sound scary to our hypervigilant amygdala, but farmers have long created mutations, originally through crossbreeding and, after Gregor Mendel, the founder of genetic science, through outright genetic manipulation. "As we began to understand how genetics worked," Diamandis and Kotler explain, "scientists tried all kinds of wild techniques to induce mutations. We dipped seeds in carcinogens and bombarded them with radiation, occasionally inside of nuclear reactors. There are over 2,250 of these mutants around; most of them are certified 'organic.'"[17]

Today, thanks to modern genetic engineering techniques, we don't have to irradiate food to change its genetic qualities. We can design seeds that don't require plowing or chemicals to grow, reducing soil erosion and the use of petrochemicals and herbicides. In 2002, farmers in India adopted genetically modified cotton and went immediately from being a cotton importer to a major exporter, increased their yields by 50 percent, reducing pesticide use by 50 percent, and nearly doubling their income to $1.7 billion a year.

Since the year 2000, the cost of sequencing plant genomes, which then took seven years and $70 million, now costs about a hundred bucks and takes three minutes. Today we can affordably modify food to better grow in hot, dry, drought-like conditions, or add nutrients to high-yield rice that can prevent children from going blind from a lack of Vitamin A, or add the omega-3 fatty acids we can currently only get from fatty foods that lead to heart disease. All this can be done while reducing greenhouse gas emissions and not destroying our soil with unnecessary chemicals.

I'm not suggesting we should move forward without caution. But I am saying we can't let our fears prevent us from moving forward at all. In 2012, I watched an astonishing video of a giant 3-D printer, created at the University of Southern California, printing a full-sized house. This process is now revolutionizing home building and is doing so with fewer resources at a fraction of traditional costs. The technology was invented to provide clean and affordable housing to poor people living in shantytowns across the globe, but it's poised to become the main method of construction for everything, everywhere. In 2019, a company in China completed the largest 3D-printed structure in the world, a 500-meter long retaining wall in Suzhou. The relatively inexpensive project was designed to preserve coastal habitats without upsetting them or polluting their waterways during its construction.

It's estimated that nearly half the world lacks the sanitation necessary for good health, which leads to the deaths of about 100,000 children a year. But even the developed world is still using 19th century plumbing technology that requires enormous wastewater treatment facilities because it remains necessary to flush human excrement into the same water people drink and wash with. In response, the Bill and Melinda Gates Foundation has invested hundreds of millions of dollars developing waterless toilets that use chemicals to turn human waste into usable compost, that don't attract mosquitos, and aren't connected to a sewer system. The Foundation's $350 Tiger Toilet, that uses worms to quickly turn refuse into compost, is already being used by thousands of the world's poorest citizens.

Bill Gates is among a growing number of modern techno-philanthropists willing to contribute much of their fortune to solving problems plaguing us all. The book, *What Are You Optimistic About?* is a collection of essays by some of today's leading thinkers, including Steven Pinker, Brian Green, Jared Diamond, Richard Dawkins, and many others, who are optimistic that violence is declining, war will come to an end, world peace is going to be achieved, our energy challenges will be sustainably met, reason and science will soon triumph, global equality will be achieved, cancer will be cured, altruism will abound, diversity will be celebrated, and so much more. Randolph Nesse, a professor of psychology at the University of Michigan, says, "I am optimistic we will soon find effective new methods for blocking pessimism."[18]

I'm not so sure I want to block my pessimism as much as I want to make certain my pessimism doesn't unnecessarily block me from moving forward. Perhaps it is also Bill Gates who provides us the best insight into finding a balance between fear and progress. On his personal blog, Gatesnotes.com, he offers a review of *The Rational Optimist*. He praises the book for its strengths, particularly for the importance Ridley places on the significance of trade in human progress, and for taking on pessimism in general. But Gates goes on to ask, "Is his optimism justified because things always just happen to work out? Or do good results depend partly on our caring and taking action to prevent and solve problems? These are important questions, and he doesn't answer them."[19]

I agree with this critique. Optimism can make us feel as apathetic about addressing our problems as pessimism can. But the world isn't getting better on its own. Slavery, racism, sexism, and so many other injustices have only been slowly improving because of people still working and struggling to make a difference. The birthrate may be leveling off, as Ridley suggests, but this is partly because of those who have been working to educate others about contraception and to raise the living standards and increase opportunities for women around the world. It may well be that Bangladesh has halved its birthrate, but it remains one of the most impoverished and overcrowded places on Earth. The many positive trends are good news, but they don't just happen. We all still have plenty of work to do.

As Bill Gates says, "Worry about fewer things while understanding the lessons of the past, including lessons about the importance of innovation."[20] To which I might add my own strategy. We need pessimism to remind us that things still need to change, and optimism to help us believe we can change them. We can't let worry paralyze us from making progress and contentment impede us from taking action. Let's move forward, ever forward, remaining cautiously optimistic.

---

[1] Ridley, Matt, *The Rational Optimist*, Harper Collins Publishers, New York, NY, 2010, p. 206.
[2] Ibid., p. 205.
[3] Ibid., p. 204.
[4] Ibid.
[5] Ibid.
[6] Ibid., p. 205.
[7] http://www.guardian.co.uk/commentisfree/cif-green/2010/jun/18/matt-ridley-rational-optimist-errors
[8] Diamandis, Peter H., & Kotler, Steven, *Abundance*, Free Press, New York, NY, 2012, p. 32.
[9] Ibid.
[10] Armstrong, Karen, *The Battle for God: A History of Fundamentalism*, (Random House, New York, NY, 2000. 2001) p. xiii.
[11] Strozier, Charles B., Terman, David, M., Jones, James W., & Boyd, Katherine A., *The Fundamentalist Mindset*, (Oxford University Press, New York, NY, 2010) p. 11.
[12] Ibid., p. 81.
[13] Ibid.
[14] Ibid., p. 84.
[15] Ibid., p. 49.

[16] See, "The Social Psychology of Humiliation and Revenge," Bettina Muenster and David Lotto, *The Fundamentalist Mindset*, ibid., p. 77.
[17] Diamandis, Peter H., & Kotler, ibid., p. 103.
[18] Brockman, John, ed., *What Are You Optimistic About,* Harper Collins, New York, NY, 2007, p. 318.
[19] http://www.thegatesnotes.com/Books/Development/Africa-Needs-Aid-Not-Flawed-Theories
[20] Ibid.

NINE

———

THE EVOLUTION OF ONENESS

In 1971, in what many consider one of the greatest songs ever written, John Lennon invited us to *Imagine* a world with none of the abstractions that divide us, including "no countries." As unthinkable as his profound vision remains, Lennon refused to be dismissed as a mere dreamer, saying, "I'm not the only one. I hope someday you'll join us, and the world will live as one."

Today, nearly half a century later, his vision of a world living as one, without countries, without religion, without possessions, or anything else to "kill or die for," still seems like a dream far beyond our reach. But I would argue it's not only within reach, it is inevitable. It may take a while before we all learn to live in peace, but the end of nations is already nigh, thrust upon us by the realities of 7.5 billion technologically advanced people living on a small planet in which empires and nations no longer work or make sense. The old notion that our greatest challenges stop at our imaginary borders—even if they are reinforced with guards, guns, and walls—is preposterous in our era of global economics, global employment, global business, global communications, as well as global poverty, global inequality, global warming, and global pandemics.

In his book, *One World: The Ethics of Globalization*, moral philosopher Peter Singer says, "Until recently such thoughts have been the dreams of idealists, devoid of practical impact on the hard realities of a world of nation-states. But now we are beginning to live in a global community."[1] Some of the evidence for this new beginning, Singer says, includes the World Trade Organization, the World Bank, the International Monetary Fund, and the International Criminal Court, along with international agreements like NAFTA, the KYOTO Protocol limiting greenhouse gas emissions, and, as of December 2015, the Paris Climate Agreement, representing an international response to global warming (though officially abandoned by the U.S. after the election of Donald Trump).

Obviously, some of these global institutions and agreements don't always go far enough and some can do more harm than good. But we are increasingly finding it necessary to govern ourselves and conduct our business as one large global community, rather than as sovereign nations able to do whatever we want. As historian Yuval Harari says in his bestselling book, *Sapiens: A Brief History of Humankind*:

> Not one of them is really able to execute independent economic policies, to declare or wage wars as it pleases, or even to run its own internal affairs as it sees fit. States are increasingly open to the machinations of global markets, to the interference of global companies and NGOs, and to the interference of global public opinion and the international judicial system. States are obliged to conform to global standards of financial behaviour, environmental policy and justice. Immensely powerful currents of capital, labour and information turn and shape the world, with a growing disregard for the borders and opinions of states.[2]

Even though our old provincial mental constructs haven't yet caught up with this new global reality, we are already increasingly functioning as one world. As Harari says, "As the twenty-first century unfolds, nationalism is fast losing ground."[3] Of course there are many, including certain politicians and leaders, desperate to maintain the status quo by securing our imaginary borders and keeping the old battle lines drawn. But such desperation already makes them appear more foolish than sensible in the wake of an inevitable reality that is already emerging.

It is inevitable because globalization is a force of nature, a consequence of evolution, and we are currently near the tail end of a process that has been happening since the dawn of human history. In addition to biology, our technologies, cultures, societies, even our ideas evolve, which means they also converge, becoming exponentially more complex and more unified in the process. Remember what Ray Kurzweil says, "Exponential growth is deceptive. It starts out almost imperceptibly and then explodes with unexpected fury."[4]

When applied to human civilization, this principle suggests the number of nations will become fewer at an exponential rate until we begin behaving and seeing ourselves as one unified species. This is precisely what is now happening. But to see it, Harari says even a bird's eye view is too narrow. He says our perspective needs to be more like a distant spy satellite, viewing history in terms of millennia, not centuries. "From such a vantage point it becomes crystal clear that history is moving relentlessly towards unity."[5]

Calculating our trajectory based upon how things evolve in general allows us to understand where our world is heading, and it's heading toward greater unity at a faster rate than ever before.

"Over the millennia, small simple cultures gradually coalesce into bigger and more complex civilizations, so that the world contains fewer and fewer mega-cultures, each of which is bigger and more complex."[6] Harari goes on to point out that 12,000 years ago there were many thousands of different human communities living on Earth, few of which knew of each other. Two thousand years ago, the number of these "dwarf worlds," as he calls them, had shrunk to just hundreds, or maybe a couple of thousand at most. Five-hundred years ago, "90 percent of humans lived in a single mega-world: the world of Afro-Asia."[7] The other ten percent lived in but four distinct societies: the Mesoamerican, the Andean, the Australian, and the Oceanic worlds. "Today," he says, "almost all humans share the same geopolitical system ... the same economic system ... the same legal system; and the same scientific system."[8]

Yet we still seem far from establishing the peaceful world John Lennon asked us to imagine, a world with nothing to kill or die for. There is still plenty of killing and dying going on, as well as lots of greed, poverty, inequality, and injustice.

And some of the global institutions and agreements we've established have only worsened the plight of many people. But there is also reason to believe our continued advance toward world unity will eventually prioritize our shared concerns. Combating global warming, economic inequality, overpopulation, pandemics, and war will begin to take precedent over the systems of greed and injustice a few are still trying to prop up and expand in this new age.

So it's important that we begin speaking in terms of our new reality, openly speaking of the end of nations, at least as we have known them. By acknowledging this new reality, we can better and more honestly address it and work together to erect new institutions and new agreements that consider the needs of everyone in the world. Many years ago, Erich Fromm explained that "equality" doesn't mean "sameness," it means, "oneness."[9] Our new global community doesn't mean we're all going to get along because we're just alike, with the same beliefs and the same ways of doing things. What it means is we should all be treated as one people despite our differences, that we should have one set of rights and be treated as equals under the law, with the same privileges and opportunities no matter who we are, what we believe, or where we are living.

A good starting point for this new reality is a return to and reinforcement of the Universal Declaration of Human

Rights established by the United Nations in 1948. It needs to be tweaked and modernized to explicitly include global protections for gays, lesbians, and transgender people, as well as securing our rights to a healthy environment. And even the concept of the United Nations itself has become inadequate under the new reality. As one global community we need an institution that represents more than the interests of obsolete Nations. It must include those belonging to the United Peoples of the world, people who are no longer bound by borders or defined by nationalism. A global community must be about the sovereignty of persons, not nations, and individuals must begin to identify more with our common humanity than our nationalities. We are one.

---

[1] Singer, Peter, *One World: The Ethics of Globalization*, 2nd ed., Yale University Press, U.S. 2004, p. 196.
[2] Harari, Yuval Noah, *Sapiens: A Brief History of Humankind*, Harper Collins Publishers, New York, NY, 2015, (Kindle version), loc. 3201.
[3] Ibid. loc. 3185
[4] Kurzweil, Ray, *The Singularity is Near*, Viking, Penguin Group, New York, NY, 2005, p. 8.
[5] Harari, ibid., loc. 2562.
[6] Ibid.
[7] Ibid. loc., 2607.
[8] Ibid.
[9] Fromm, Erich, *The Art of Loving*, (Harper & Row, New York, NY, 1956), 12.

PART TWO
_____

POST-SU

TEN

THE SINGULARITY AND ME

In 2018 I attended a weeklong course at Singularity University in Moffett Field, California, on the NASA Research Park campus, sitting among more than 90 company executives and entrepreneurs from 34 different countries, listening to lectures from industry experts about artificial intelligence, virtual reality, gene editing, the microbiome, brain implants, autonomous cars, and other innovative technologies that are already radically altering our world. I admit, it was an odd environment for a minister to invest his time and resources in. Silicon Valley isn't exactly Mecca, but it called to me as sure as the Camino de Santiago, Stonehenge, or Machu Picchu calls to other sojourners.

When some of the other participants expressed friendly curiosity about my presence, most of whom were from other countries and had never heard of my Unitarian Universalist religion, it was a challenge to dismantle their misconceptions about what being a "minister" means to begin with, to introduce a little about how mine differs from other faith traditions, and to explain my reasons for being there, all in just a few introductory comments.

"My faith embraces science and reason as a source of inspiration," I'd say, "and, in my studies, I've long understood evolution to be a process of convergence through which everything is moving towards unification and greater states of consciousness." This is what Pierre Teilhard de Chardin, who helped discover the Peking Man, was getting at early in the 20th century when he said, "Life moves toward unification."[1]

The founders of Singularity University, Peter Diamandis and Ray Kurzweil, first encroached upon my professional turf—Diamandis with his inspirational message of hope for the near future, and Kurzweil by sometimes using theological language to describe how technology will enable the Universe to "wake up" and become "sublimely intelligent." "That's about as close to God as I can imagine,"[2] he says.

THE SINGULARITY AND ME

I first heard about Singularity University in 2012 while reading Diamandis's book, *Abundance*. "Each year," he explains, "the graduate students are challenged to develop a company, product, or organization that will positively affect the lives of a billion people within ten years."[3] This challenge, I learned at SU, is referred to as a student's 10x Vision. My vision, it may not surprise you, is to harness exponential technologies to help bring the global human family together. I envision, for example, a platform that allows people all over the world to have a voice and vote on the issues impacting us all, or using virtual and augmented reality to help some of the poorest or most isolated people anywhere—behind a wall in Gaza, impoverished in Sub-Saharan Africa, locked away in prison—to find employment anywhere on the planet without leaving where they are, and using AI to help us all instantly transcend our language barriers. If this sounds like an impossible dream, keep in mind the technologies for making it happen already exist.

Let me back up by talking about the meaning of the terms, *exponential* and *the Singularity*. In 2001, Ray Kurzweil wrote an essay, "The Law of Accelerating Returns," in which he distinguishes between "intuitive linear" and the "historical exponential" views of technological progress. In brief, he explains, people "intuitively assume that the current rate of progress will continue for future periods,"[4] that things will evolve at the same rate we're experiencing now. But from the broader historical view, he says, "Exponential growth is a feature of any evolutionary process, of which technology is a primary example."[5] Computational power, for example, doubles every year. Its evolution isn't a steady linear line upward, but a line that sharply bends from a straight horizontal incline to a sudden vertical line going almost straight up. I won't go any further into the theory here. All you need to do is look at the technological advancements wrought during the past hundred years, which have especially accelerated during the last few decades, and you'll understand the meaning of exponential change.

In 2005, Kurzweil published his bestselling book, *The Singularity is Near*, referring to the point at which technology allows intelligence to transcend biology, leading to an explosion of rapid changes that will make all the advances made during the next 100 years look "more like 20,000 years of progress"[6] at today's rate. During my week at SU, Pascal Finette, its Chair for Entrepreneurship & Open Innovation, pointed out that if we had a time machine that could bring Jane Austen 200 years into the present, "she'd probably go crazy because of all the changes." But if we brought someone from the 15th century 200 years forward into Jane Austen's time, they would notice few changes, if any. "To get someone to have the same, mind-blowing experience she had visiting us," Finette says, "she'd have to bring someone from 13,000 BC," the Stone Age, before agriculture began. Diamandis calls this kind of reaction "disruption," and you can see it on YouTube watching the responses of people who experience virtual reality for the first time, or sitting in an autonomous car that suddenly speeds up to 75 miles per hour with nobody driving.

In a world of exponential change, amidst a technological singularity, humans will have to live in a world of constant disruption that requires us to continuously learn and adapt—a world that has already begun. Driverless cars are a good example. On my way to the airport after my week at SU had ended, I asked my Uber driver what he was going to do when autonomous cars take over the roads. He insisted it won't happen for a very long time and isn't something he needed to worry about any time soon. Maybe, but consider this: during another lecture, Neil Jacobstein, SU's AI and Robotics Chair, showed a 1904 photo of a busy New York City intersection filled with horses and only two cars. A 1917 photo of the same congested intersection shows no horses and all cars. In 1900 there were only 8 thousand cars in the world. Just twenty years later, there were 8 million, and that was at a time technology was creeping along compared to the present.

General Motors Chairman Mary Barra has predicted that the auto industry will change more in the next five years than it has during the past fifty, and says automated cars will soon become a "second office," allowing passengers to work while AI does the driving. It also means those who are no longer able to drive themselves don't have to give up their autonomy. They can still go anywhere they want at any time. The technology is already here. Now it's just a matter of waiting for mass production and regulations to catch up.

I also learned that between 2015 and 2016 two-legged robots from a variety of engineering groups went from constantly and comically losing their balance and falling over while conducting difficult tasks and traversing uneven terrain, to being able to run upstairs and do back flips thanks to advances in Artificial Intelligence—again, in just one year. There are also robots so small they can stand on the cross-section of a human hair, and nano-sized robots powered by electromagnetic fields that can be ganged together, like the individual cells of a muscle, to perform tasks. Although AI isn't self-aware just yet, it is now able to make decisions on its own, thanks to deep learning: that is, by collating information from thousands, even millions, of sources in an instant. AI robots are already making decisions on the surface of Mars, beneath arctic ice sheets, on our roadways, and increasingly in the realms of medicine and criminal law.

Suzanne Gildert is founder and CEO of Sanctuary AI, a company working to build embodied AI robots that are indistinguishable from humans. She reasons that if we want AI to have our values and be able to relate to us and us to them, they need to learn as humans, with humans, in human environments. She doesn't worry about them being heartless machines. On the contrary, she expects them to obtain hyper-empathy and compassion and to serve as our companions, therapists, caregivers, nurses, and even our entertainers. Maybe you've seen Sophia, the humanoid robot developed by Hanson Robotics, making the rounds on late-night television in recent years. She's the first robot to have been granted legal citizenship, in this case by Saudi Arabia.

Impressive as they are, these AI robots are far from being indistinguishable from human beings. "We have a demo model out at the university interacting with people," Gildert says. "That doesn't feel like it's moving very fast but compare it to how many Teslas were shipping in 2008 versus today." Tesla delivered 321 cars four years after building its first Roadster in 2008. In 2018 it delivered more than 80,000. That's 26-thousand percent exponential growth.

I think companies like Sanctuary AI will succeed in creating humanoid robots sooner than most of us imagine, but I also think they will be obsolete before they get here. They'll be an unnecessary novelty due to the much more rapid advancement of Extended Reality, meaning virtual and augmented reality. Virtual reality, which hit the consumer market only a short time ago in 2015, is already capable of transporting us anywhere in the Universe, to a precise replication of Mars, for instance, developed by NASA, based on Rover and other lander photos.

For the first time, topographers can walk on virtual simulations of Mars to accurately map its surface. We can visit with others from around the world in virtual social spaces and meeting rooms. We can go into a virtual submarine and explore the sunken Titanic, visit museums, climb Mt. Everest, fly to other planets, relax on the beach, base jump, rock climb, go to a large-screen theatre in the comfort of our own homes, and so much more.

Even more importantly, virtual reality is helping us rewire human brains to bypass damaged bodies where certain connections have been lost. Paraplegics are regaining the ability to move their legs by learning to walk in VR. Burn victims living in chronic pain alleviate their anguish by going into a frozen VR environment and throwing snowballs at virtual penguins. It's already being used to train athletes in professional sports, to take schoolkids on virtual field trips, to train employees at Walmart and Lowe's, and to learn how to operate heavy equipment without worrying about dangerous or expensive accidents.

Augmented reality doesn't create a virtual space, but virtual objects in real space. Instead of owning a physical television, or a computer, or even a cellphone, we can simply put on a pair of glasses or contact lenses and have all these things appear right in front of us, in whatever environment we choose to be in. The dematerialization of stuff is going to save lots of resources. Being able to go anywhere and be with anyone will also reduce the need to travel and burn fossil fuels. Dematerialization will likewise allow AI to exist anywhere, in whatever form we want, as a person, a talking dog, what have you, or whatever form it chooses for itself.

I don't see any reason for AI to get locked into a physical body. There may be robot bodies any of us, AI or human, can temporarily inhabit and control through a VR headset whenever we need to have a physical presence elsewhere—to do a dangerous job, for example—but I don't see a widespread need for AI to have physical bodies. Since Extended Reality is already so advanced, I just don't see a massive need for embodied AI.

Speaking of going to space, most everyone has heard that humans are planning to go to Mars by 2030. The idea is that we will someday inhabit it, as a "Plan B" in case Earth doesn't work out. Even though the trip to Mars is likely to happen and some may eventually live there, many experts don't consider it a viable plan for expanding our presence beyond Earth. I was surprised to learn, upon hearing Gregg Maryniak at SU, an internationally recognized NASA scientist, that the idea of constructing cities in space is being taken far more seriously by many than is finding another planet to live on. Plans already exist for space cities the size of Switzerland that would include gravity and shield their inhabitants from cosmic radiation. These cities will dwell in what's called "free space," outside the gravity wells planets are in. Escaping from the pull of these wells is the most dangerous part of space travel, which isn't an issue in free space. The billionaire Space Baron Jeff Bezos has speculated a trillion of us will eventually live in these space cities, far more than will ever inhabit the Earth. Again, these plans are already in the works and billions of dollars are being spent by entrepreneurs like Bezos to help make them happen.

More exciting than space exploration, however, is the positive impact technologies like these are already having on our environment, particularly in response to global warming.

There are five major sources of carbon emissions—electricity, agriculture, manufacturing, transportation, and buildings, none of which is a main culprit, and all of which are being impacted by emerging technologies. For instance, we are beginning to use largescale 3-D printers to construct affordable, energy-efficient houses, using very few materials, at a cost as low as $4,000 each. Not only does this save resources, it's how ordinary communities can address homelessness and, perhaps, lead to a world that can afford to guarantee housing as a basic human right.

3-D printing is also being used to make parts for things like airplanes and automobiles that are lightweight and use fewer resources than traditional manufacturing. In agriculture, cows produce about as much CO_2 as energy production because of the methane gas they generate. Today, a variety of companies are growing real meat and leather out of stem cells in laboratories. The process doesn't kill animals, is cost effective, and uses a lot less land and water compared to live animal production. Humane meats are expected to be on the market in the very near future.

You may have heard that Peabody Coal went from an all-time financial high in 2011 to bankruptcy just two years later. In fact, between 2011 and 2017, 75 percent of coal companies around the world went bankrupt. Natural gas is one reason, but so is the growing renewable energy industry. Since 1980 the cost of wind energy has plummeted and is now selling for as little as 2 cents per kilowatt, the lowest price for energy ever. A big part of this is the growth of wind production driving the cost down, along with exponentially improving wind-harnessing technologies. Likewise, since 1980, the cost of solar energy has dropped 250 times, from 77 dollars per kilowatt, to just 3 cents. Today, the cheapest coal and oil produced energy is twice that amount. It's predicted solar energy will cost a penny per kilowatt in the 2020s, an industry that grew 50 times larger between 2008 and 2018. Since wind and solar energy can be combined, and one is produced better at night and the other during the day, and because of new battery technologies for storing these energies, they are expected to soon cover most of the world's energy needs.

Of course, the damage is already done: the climate has already changed, even if we dramatically reduce our carbon emissions. This means we need to develop technologies to help reverse its impacts. Fortunately, this is happening too. Technologies now exist that can remove carbon from the air. Drones are being used to go to remote areas where they can impregnate the ground with new trees. In 2018 a team met the XPRIZE Foundation's challenge to develop a technology that can collect "a minimum of 2,000 liters of water per day from the atmosphere using 100 percent renewable energy, at a cost of no more than 2 cents per liter." Skysource/Skywater Alliance, the winner of the challenge, is already using this technology to pull fresh water out of thin air in some of the most parched places on the planet.

There are many other transformative technologies I learned about at SU that I won't elaborate on, but include gene editing that already allows us to turn off those genes responsible for things like Alzheimer's, HIV, and certain types of cancer; along with memory-enhancing microchips. Nor will I go into some of the frightening possibilities some of these same technologies could lead to, like AI taking over the world, the weaponization of gene editing, the misuse of our data, and so forth. There are sure to be unforeseen consequences. A future without problems is about the only thing I can guarantee will never happen.

I'm not preaching a techno-utopia here, but I am personally inspired by all I learned at SU because it gives me tremendous hope for both today and tomorrow at what often feels like one of the most daunting moments in human history. We face some of our greatest challenges ever, yet also have some of the most powerful solutions ever. When President Bill Clinton met Peter Diamandis, the author of his favorite book, *Abundance*, Clinton asked, "Why are you so positive? Don't you watch the news?"

"No," Diamandis told him, not because he tries to avoid disturbing information but because he doesn't believe it's true. During his lecture at SU, Diamandis pointed out that bad news is reported 10 times more than positive news, and it almost entirely misses all the progress we're making. As a result, most of us don't realize things like income, lifespan, and food availability are up, while the costs of energy, transportation, and communication are down. In 1820, 94 percent of the world's population lived in extreme poverty. Today it's 10 percent. Back then, 83 percent had no education. Today 86 percent have a basic education. Then, 88 percent were illiterate. Today it's only 15 percent. Only 1 percent lived in democracies. Today it's 56 percent. No one was vaccinated against common diseases. Today 86 percent are. In 1820, 43 percent of children died before age 5. Today it's less than 4 percent.

Before attending Singularity University, I didn't know what my pilgrimage to Silicon Valley would yield. But the hope and optimism I felt reading the work of its cofounders, Peter Diamandis and Ray Kurzweil, had been pulling me toward this unorthodox sanctum ever since I first heard of it. I wasn't disappointed. I left SU transformed. Not only were my reasons for hope and optimism soundly validated, but I became part of a community of extraordinary people with the wit, will, passion, and resources to transform our world and our lives for the better. I went seeking hope for tomorrow and left with the promise of today. I went as an American who sat among nearly a hundred strangers from countries around the globe and became part of one human family that transcends nationality. I went searching for a way to make a difference and became part of a movement that's already making a difference.

---

[1] de Chardin, Pierre Teilhard, *The Future of Man*, Harper & Row, New York, NY, 1959, 1964, p. 72.
[2] Kurzweil, Ray, *The Singularity is Near*, Viking Press, New York, NY, 2005, p. 389.
[3] Diamandis, Peter H., & Kotler, Steven, *Abundance: The Future is Better than You Think*, Free Press, New York, NY, 2012, p. 73.
[4] http://www.kurzweilai.net/the-law-of-accelerating-returns
[5] Ibid.
[6] Ibid.

ELEVEN

CHANNELING YOUR INNER THERAPIST

Two books have had a profound impact on my thinking, books that I've never read. That's because their titles are so good, they say it all. The first is James Hillman's and Michael Ventura's 1993 collaboration, *We've Had 100 Years of Psychotherapy and the World's Getting Worse*. It's now been more than 125 years and things haven't gotten much better, at least not in the realm addressed by psychotherapy, psychology, and psychiatry, which I'll summarize as peace of mind, or contentment (*ataraxia*, as the Greek philosophers called it). We all want to be happy with our lives, with ourselves, with the way things are, but few of us are. Although it's a relative term, Hillman and Ventura say things are *worse* than ever, from which they seem to infer psychotherapy isn't working.

In a more recent book, *The Coddling of the American Mind*, which I have read and recommend others read, Greg Lukianoff and Jonathan Haidt confirm that rates of suicide, depression, and anxiety are greater than ever among today's newest generation, iGen: those born in the Internet age starting around 1995. In the U.S., their data indicates the percentage of iGen college students describing themselves as having a mental disorder, mostly anxiety and depression, has "increased from 2.7 to 6.1 for male college students between 2012 and 2016 (that's an increase of 126%). For female college students, it rose even more, from 5.8 to 14.5 (an increase of 150%)."[1] Anxiety is now the "leading problem for which college students seek treatment," they say, further pointing out that during this same period there have been "substantial increases in rates of self-injury and suicide among college students."[2] According to this data, at least when it comes to peace of mind, things really do seem to be *worse* than ever.

But *worse* is a relative term. There's also lots of data indicating the world today is better than ever—*better* being another relative term. According to numerous sources including *Abundance* and *Bold*, both coauthored by Peter Diamandis and Steven Kotler; *The Rational Optimist*, by Matt Ridley; the anthology, *What Are You Optimistic About?*, edited by John Brockman; *Factfulness*, by Hans Rosling; and numerous reports you can get right off the United Nation's website—things are better than many of us think.

Take the most recent of these, Rosling's 2018 bestseller, *Factfulness*.[3] According to the data Rosling says that during the last 20 years alone the number of people in the world living in extreme poverty has been cut in half; the majority of the world's people now live in middle-income countries; 60 percent of girls living in low-income countries complete primary education; the number of people being born in the world has leveled off, and the number of children today is expected to be the same number of children in 2100; the global population will continue to rise until then, by another 4 billion people, before leveling off, mostly because people are living longer; the average life expectancy round the world is 70 years; during the past century, the number of us killed in natural disasters has been halved; 80 percent of the world's children have been vaccinated against common childhood diseases; by age 30, men have spent an average of 10 years in school, compared to 9 years for women the same age.

Rosling and others leave us with lots of verifiable statistics like these. Yet his book, in particular, also includes surveys about what most people think is true about these matters. In all categories, most people think things are far worse than they are. In fact, only 10 percent of the 12,000 people Rosling tested from 14 countries scored as well as a group of chimpanzees would have at random: they would have at least gotten it right about 33 percent of the time. Most people answered only 2 out of 12 questions correctly.

Part of the problem, from the standpoint of evolutionary psychology, may be that humans aren't wired to think positively about the future. Ridley says "a million years of natural selection shaped human nature to be ambitious to rear successful children, not to settle for contentment: people are programmed to desire, not to appreciate."[4] Citing Ridley in *Abundance*, Diamandis and Kotler more succinctly say, "We might be gloomy because gloomy people managed to avoid getting eaten by lions in the Pleistocene."[5] Contented people, in short, didn't survive evolution as well as their discontented counterparts.

If this is so, if most of us are "gloomy" by nature, because of natural selection, it may explain why, after 125 years of studying how the human psyche works, psychotherapy hasn't helped us improve much. How can we cure discontentment if we have evolved to be discontented? Or maybe Hillman and Ventura themselves are glumly looking at the glass half empty. Maybe more of us are happier today than we think. Or maybe things are better than ever today because lots of people worried about them to begin with. How would we ever progress if we were all content with the status quo, with how things are, with good enough? Maybe a little unhappiness in the world isn't such a bad thing.

I have studied psychology as a means of personal growth and wellbeing more than most novices. I even wrote my doctoral dissertation on the subject. Yet I agree, psychotherapy hasn't solved all our problems, nor should it be expected to. Since its advent more than a century ago, the world has gotten worse in some ways and better in others. But this isn't reason to infer things have gotten better or worse because of psychotherapy. Psychotherapy is one tool in our toolbox, but it isn't a tool that can fix everything in our lives or solve all the problems of the world. The word *therapy* means, "to assist," and that's all it's meant to do, to assist us in our own work as individuals. But we still have to take responsibility for our own lives, growth, and wellbeing.

The other book I've never read that's transformed my outlook on life is *If You Meet the Buddha on the Road, Kill Him! The Pilgrimage of Psychotherapy Patients*, by Sheldon

B. Kopp. The full cover actually says, "No meaning that comes from outside ourselves is real. The Buddhahood of each of us has already been obtained. We need only recognize it. Thus the Zen Master warns his disciple: If you meet the Buddha on the road, kill him!"[6] As a psychotherapist himself, which he views as a type of modern guru, Kopp isn't encouraging animosity toward therapists, only that each of us must take responsibility for our own development and welfare. A good therapist can help point us in the right direction, or, even better, help us discover the right direction for ourselves, but the journey, with its unexpected turns, up and downs, and rough going is ours to take.

I would also add, as an empiricist, I don't agree that "no meaning that comes from outside ourselves is real." I think one of the problems with our world today is far too many of us are unwilling to incorporate factual data and hard evidence into our worldviews. Whether it's fact-free conservatives denying global warming or postmodern liberals obsessed with the right to believe whatever they want, we'd all be better off if more of us embraced the hard realities outside our own heads. The inability to allow the empirical world to influence us is called *autism*, in the classic sense. Still, I do like Kopp's point about killing the Buddha, which serves to remind me I must become my own guru, the person most responsible for my own unfolding and health.

So this is the theory of psychotherapy I've adopted in my life, based upon the titles of these two books I've never read: that things can become better or worse with or without psychotherapy, which should be seen as a tool for aiding us in our own intrapersonal work; and that we must each ultimately take personal responsibility for our own growth, fulfillment, and happiness. My guess is most psychotherapists, psychologists, and psychiatrists would agree.

There are a few practices I've found especially helpful, the first of which is *attitude*. Evolutionary Psychology suggests we may naturally be inclined toward a pessimistic outlook, even if things aren't as bad as we think. This really hit home for me when I was studying the psychology of fundamentalism while working on my dissertation and learned one of its earmarks is *apocalypticism*, the unfounded fear that things are so bad, that one's problems are so great, the whole world is ending. It's a narcissistic projection and exaggeration of one's fear instinct, which under normal circumstances can keep us safe but should subside once we've assessed a situation and determined there aren't any real threats. But for some, the state of hypervigilance our amygdala is prone toward is always on alert, leading to a life of unwarranted anxiety, paranoia, conspiracy theories, and the general belief or feeling that things are worse than ever and only getting worse.

But you don't have to be a conservative fundamentalist for this to be so. In *The Coddling of the American Mind*, Lukianoff and Haidt point out it's often progressives who succumb to what they call *catastrophizing*, which they define as "Focusing on the worst possible outcome and seeing it as most likely."[7] I realized, several years ago, this had been true of my own thinking, that I was prone to seeing and fearing the worst. Simply becoming aware of this tendency, however, has changed my life and my outlook. I now have a much more positive view of the world, hope for our future, and far fewer dragons to slay or windmills to topple. By simply becoming aware of this cognitive distortion in my own life, I've been able to keep it in check and be a happier person.

This leads to the second practice, researching and accepting the facts. In days past, this was a lot harder to do. It took a lot of time to find and learn the facts, and those who did were considered experts in their fields. Today the Internet has made it easy for most of us to find reliable data-driven reports from reputable sources. Looking at the real numbers can help us reevaluate our unsubstantiated presumptions. The research is easy these days but finding the will to challenge our own beliefs is no easier than it's ever been. Many would prefer to simply get their information from Fox News or MSNBC; or base it upon the scrutiny and acceptance of one's own peers on social media. But to be healthy, happy individuals, it's necessary to be honest with ourselves and embrace the world on its own terms. As Lukianoff and Haidt explain, "If you can get people to examine these beliefs and consider counterevidence, it gives them at least some moments of relief from negative emotions, and if you release them from negative emotions, they become more open to questioning their negative beliefs."[8]

A third practice is logically evaluating what we hear. This isn't as easy as using the Internet because most of us have never learned to reason, though most of us like to think of ourselves as perfectly reasonable. Reason is a discipline that must be learned and practiced, just like trigonometry or playing the saxophone, neither of which comes naturally. What does come naturally is thinking that justifies our emotions, that justifies what we *want* to be true. These "Rationalizations," social psychologist Erich Fromm says, "are essentially lacking this quality of discovering and uncovering; they only confirm the emotional prejudice existing in oneself."[9] Your inner therapist can get into the habit of asking if your thinking is helping you learn something new, or if it's just confirmation bias. If our own minds are mostly in agreement with what we already believe, we're probably not being as reasonable as we like to think.

Learn logic if you can, but if you can't, there's something much easier you can do to make yourself a better thinker. Get back in the habit of asking *why?* I say, "get back in the habit," because we begin our lives by incessantly asking *why* about everything, until we hear, "because I said so," or, "don't ask questions," or, "never question the authorities," one too many times. But logic is the study of what follows, of making inferences, of figuring out *why* something is said to be true. So train yourself to let the questions start coming naturally again. If somebody asserts something is true, or you catch yourself believing something is true, ask yourself *why*, and then consider if the *why* makes sense.

You'll notice all these practices so far are about adjusting our mindset, and that's what self-therapy is about: having a more positive mindset, or at least a more realistic one, by expanding our awareness, recognizing our tendency—the human tendency—to expect the worst, looking at empirical data, learning logic, and asking *why?* This is because self-therapy is more about changing our minds than our circumstances.

It's with this in mind that I also recommend bibliotherapy, a technique I learned from philosopher Lou Marinoff, author of, *Plato, Not Prozac*, and founder of The American Philosophical Practitioners Association. In discussing philosophy with his clients, in the hope of providing them with new perspectives, Marinoff prescribes books to those who want to learn more about a helpful idea. Quoting Thoreau, he says, "How many a man has dated a new era in his life from the reading of a book."[10] Of course, bookstore shelves are filled with self-help books, but when it comes to evidence-based information the science shelves might be just as helpful, along with my favorite, the *philosophy* shelves, a discipline that requires us to question our own mindsets, the ones we're trying to change. If you're looking for a recommendation, *Plato, Not Prozac* is a good start.

These days I'd also recommend learning about the specific philosophy of *Stoicism*, which has been around 2,300 years and is now making a comeback. Stoicism is about maintaining our peace of mind and being in good spirits (*eudaimonia*) regardless of what's going on, by distinguishing between what we can and can't control, then focusing on the former. I'd also suggest studying Cognitive Behavioral Therapy (CBT) and putting it into practice. CBT can especially help with the rising rate of anxiety disorders because, rather than avoiding the things we're anxious about, it helps us face our fears and, in so doing, become aware of the many cognitive distortions that cause us to see the world incorrectly, like always expecting the worst. Interestingly, if you read modern books about Stoicism, their authors often compare it to CBT. And if you read books about CBT, their authors compare it to ancient Stoicism. I'd recommend putting William Irvine's *A Guide to the Good Life: The Ancient Art of Stoic Joy*, and the aforementioned, *The Coddling of the American Mind* on your bibliotherapeutic book list.

Another practice is making time to just relax and enjoy life, without worrying about doing anything to fix yourself or the world. Let it all go and just have fun. "Sweet, sweet surrender," as John Denver sings. Let yourself be imperfect, incomplete, and unfinished. Just be. Just take a few moments to appreciate the rare gift of being a sentient, somewhat self-aware being in the Universe. Take in your guilty pleasures with some abandon, before it's too late, because life is short (for now). As the poet Rumi said, "Drink all your passions and be a disgrace."[11]

Finally, your inner therapist needs to be someone you can trust. I don't mean you should simply trust your feelings or blindly trust your own beliefs. These can be generated from the very cognitive distortions and faulty thinking we're trying to recover from. By trusting yourself, I mean becoming a person you know does your best to see the world on its own terms by looking at empirical facts, reasoning well, asking *why*, studying, facing your fears, and taking time to rest and recreate. In short, become a person you trust by trying not to deceive yourself. Be honest with yourself. *The unexamined life is not worth living.*

This requires us to allow ourselves permission to be wrong. I don't mean just being ideologically wrong but being *ideally* wrong. It means recognizing we don't have to be perfect, and that part of being human is discontentment, a state that's always pushing us forward, to become more than we are, to make the world better, no matter how good it gets. The reason Hillman and Ventura think the world has gotten worse despite a century of psychotherapy isn't because there's something wrong with psychotherapy, but because psychotherapy isn't about fixing the world to begin with. It's good to focus on ourselves, but if we're always content, why would we ever want to make things better? To make things better we have to accept the responsibility we have toward others, to what's beyond ourselves, and to do what we can and must, even in our imperfection, to be a helpful, productive, loving presence in the world.

[1] Lukianoff, Greg & Haidt, Jonathan, *The Coddling of the American Mind*, Penguin Press, New York, NY, 2018, p. 156.
[2] Ibid., p. 157.
[3] Rosling, Hans, *Factfulness*, Flatiron Books, New York, NY, 2018.
[4] Ridley, Matt, *The Rational Optimist*, HarperCollins, New York, NY, 2010, p. 27.
[5] Diamandis, Peter, & Kotler, Steven, *Abundance,* Free Press, New York, NY, 2012, p. 39.
[6] Kopp, Sheldon B., *If You Meet the Buddha on the Road, Kill Him!*, Science and Behavior, Palo Alto, CA, 1972, Bantam Books, New York, NY, 1976.
[7] Lukianoff and Haidt, ibid., p. 38.
[8] Ibid., p. 37.
[9] Fromm, Erich, *Escape from Freedom*, (Avon Books, Heart Corporation, New York, NY, 1941, 1965) p. 219.
[10] Marinoff, Lou, *Plato Not Prozac!* HarperCollins, New York, NY, 1999, p. 165.
[11] Rumi, *Like This*, versions by Coleman Barks, Library of Congress Catalog #89-092393, 1990, p. 35.

TWELVE

MY MOONSHOT HEROES

"Moonshot" has been the title of several books about, what else, going to the Moon. It's also a term sometimes used in baseball to describe a high-flying homerun hit. But in recent years *moonshot* has come to refer to any attempt to accomplish something so extraordinary that most people would probably consider it impossible. When the original moonshot, sending an astronaut safely to the Moon and back, was proposed by President Kennedy in 1961, the technology for doing so didn't even exist and nobody had any idea how to make it happen. As moonshot champion and cheerleader, Peter Diamandis, a founder of the XPRIZE Foundation, often says, "The day before something is truly a breakthrough, it's a crazy idea." Conversely, in his recent book, *Moonshots: Creating a World of Abundance,* Naveen Jain, a successful American entrepreneur and philanthropist who grew up impoverished in India, says, "The minute you believe something is impossible, it becomes impossible for you."[1]

This new use of the term began with a relatively unknown research and development wing of Google called X, as is the Roman numeral 10. The idea of X, whether it's Google X, XPRIZE, or SpaceX, is not limiting ourselves to incremental, ten-percent-better-than-now thinking, but 10X, ten-times-better-than-now. "We start with a large problem in the world that, if solved, could improve the lives of millions or even billions of people," Google X says. "Then we propose a radical solution that sounds impossible today, almost like science fiction."[2] When Google X was just getting started, founder and CEO Larry Page was talking with Astro Teller, the lab's head, a brilliant engineer who happens to be the grandson of Edward Teller, father of the Atomic Bomb. Page mentioned he didn't like the terms, "research center" or "innovation incubator" because they're too boring. Teller thought for a moment, then asked, "So, are we taking moonshots?"[3]

"That's it," answered Page. "That's exactly what we're doing!"[4] And that's how Astro Teller got his official title, Captain of Moonshots, and why Google X is nicknamed, "The moonshot factory."

Some of the moonshots the factory has churned out so far include using balloons to send satellites into the stratosphere, to help provide affordable Internet service to the 3.5 billion people around the world who don't have it yet. X was able to use this technology to reestablish Internet service in Puerto Rico after Hurricane Maria knocked it out in 2017. WAMO, the self-driving car company, is also a product of X and is about to disrupt the auto industry by reducing car ownership, parking lots, and the number of cars on the road. X, like several other companies, is also working on autonomous flying delivery drones, which they expect will be as transformative as the Pony Express, ships, trains, and planes have been, making it possible for shut-ins to quickly get things, like dinner for the sick kids, or bringing life-saving supplies to Emergency workers in remote areas. X has also invented contact lenses capable of monitoring our health, starting with diabetes detection; kites that generate wind power without the need for colossal windmills and windfarms; salt-based batteries for storing extra solar and wind energy;

and carbon-neutral fuel made out of seawater, among other crazy ideas.

During the 2018 Abundance 360 Conference, Teller told participants the greatest obstacle to moonshots is fear of failure. "If you want your group to take moonshots," he says, "you have to start by making it socially uncomfortable for people to suggest 10 percent thinking … you want people to think that their jobs are tied to the weirdness and the bravery and the creativity."[5] Moonshot thinking means dreaming the impossible, always in the service of humanity, always "one giant leap for mankind," and having the courage to go for it no matter how crazy it might seem.

Let's say we decide to build a machine that can pull water out of thin air. Not crazy enough for you? Then let's make it no less than 2,000 liters of water a day, costing less than 2 cents per liter, using 100 percent renewable energy. Or wouldn't it be great if we had tricorders like on *Star Trek*, that can quickly scan and diagnose illnesses? What if you could have one and use it without needing to go to a doctor or using health insurance? Or how about an affordable, four-passenger car, complete with heat, air-conditioning, stereo, and plenty of room for luggage, that goes from 0 to 60 in 15 seconds and gets 100 miles per gallon—on almost anything? What if we created a way to check the pH, or acidity, of our oceans as accurately, affordably, and easily as we check the levels in a fish tank?

Crazy as these ideas sound, none of them is just an idea anymore. They have all become realities, thanks to the XPRIZE Foundation. XPRIZE was founded in 1995 for the purpose of incentivizing solutions to the world's grandest challenges. "The solutions to the world's problems won't come from one person or one country or one industry," the Foundation's website says. "We will only reach these solutions if everyone can make their voices heard."[6]

Nearly a month after BP's Deepwater Horizon explosion in 2010, spilling nearly 5 million barrels of oil, a slick covering as many as 4,000 square miles, into the Gulf of Mexico, James Cameron, director of *Titanic*, contacted Peter Diamandis at XPRIZE and suggested offering a "rapid response 'flash prize'"[7] to anyone who could figure out how to cap the oil that continued gushing for 87 days. But after studying the problem, "The idea of a prize to cap the gusher was off the table," Diamandis says. "BP would never give us or anyone access to their data."[8] So they decided to focus on cleanup instead. They found a philanthropist, Wendy Schmidt, president of the Schmidt Family Foundation, to underwrite a $1.4 million prize to anyone who could quickly develop a means of recovering the spilled oil at no less than

2,500 gallons per minute, more than twice the rate of existing technologies. Seven of the ten finalists were able to meet the challenge, but the winning team came up with a method that cleaned up almost 4,700 gallons per minute and, today, has been improved enough to clean up 6,000 gallons per minute.

That's a genuine moonshot. But the most unique team of finalists included a tattoo artist, a mechanic, and a dentist. They heard about the prize, joked about the dentist among them having the most experience with drilling, used a home jacuzzi for their experiments, and came up with a technology for sucking oil out of the ocean that beat all previous records on its first day out. Diamandis recalls tattoo artist Fred Giovannitti saying, "We get asked all the time, 'How long have you been in the oil industry?' and I ask back, 'Counting today?'"[9] That's the beauty of moonshots. All of us are capable of thinking up crazy ideas and making them come true, no matter our experience or expertise. And thanks to the quick, crazy thinking in this case, disastrous as it was, the Deepwater Horizon incident didn't become another *Exxon Valdez*, which is quite a statement considering the Horizon spewed close to 200 million gallons, compared to the 11 million spilled by the *Valdez*.

According to its website,[10] just some of the global incentive-competitions the XPRIZE Foundation sponsors or has sponsored include offering millions of dollars to teams that can "successfully advance deep sea technologies for autonomous, fast, high-resolution ocean exploration"; "develop open-source, scalable software that will enable children in developing countries to teach themselves basic reading, writing and arithmetic within 15 months"; "develop breakthrough technologies that will convert CO_2 emissions from power plants and industrial facilities into valuable products like building materials, alternative fuels and other items that we use every day"; "develop and demonstrate how humans can collaborate with powerful AI technologies to tackle the world's grand challenges"; or literal moonshots, by incentivizing "private companies landing on the Moon."

But you don't have to win an XPRIZE to make a moonshot. One of my favorite moonshot heroes is my friend Allan MacRae from New Zealand. During the 1980s, Allan was New Zealand's Youth Justice Coordinator. At the time, he says, "Thousands of children, especially members of minority groups, were being removed from their homes and placed in foster care or institutions. The juvenile justice system was overburdened and ineffective. New Zealand's incarceration rate for young people was one of the highest in the world, but its crime rate also remained high."[11]

Allan MacRae is an unusually undaunted person with the gift of seeing every problem as a possibility and every obstacle as an opportunity. After a period of listening to communities throughout the country, especially to Maori minority communities, he developed the Family Group Conference, allowing young offenders the chance to make restitution for their crimes by developing a plan they agree upon with their families and other sources of support, the police, and, most importantly, the victims of their crimes. It proved to be so successful and transformative that before the end of the decade, in 1989, the New Zealand Legislature passed the Children, Young Persons and Their Families Act, adopting this new way of dealing with juvenile justice throughout the nation. Prior to this, Allan told me, the Juvenile Court in Christchurch, the South Island's largest city, held court all day, five days a week, with the average offender standing before a judge for fewer than five minutes. Within three years of adopting the restorative justice model, it held court just half a day on Fridays with plenty of time to consider each case, and every juvenile detention center in the country was closed down. How's that for a moonshot?

In case you haven't guessed it, Peter Diamandis is also one of my moonshot heroes. Like Allan MacRae, he sees problems as possibilities: "The world's biggest problems are the world's biggest business opportunities,"[12] he says. Diamandis's goal in life, what he calls his Massively Transformative Purpose, his MTP, is to support a new breed of entrepreneur to focus on solving the world's grand challenges. It's not about the money because the money is just a byproduct of helping people. "If you want to make a billion dollars," he often says, "help a billion people." This is why he founded XPRIZE, and Singularity University, A360, and the Abundance Digital Community: to encourage the world to think big, to think crazy, to think positively, to discover our own MTPs and take moonshots to make them happen, all to create a world of abundance for everyone, every last person.

After reading *Space Barons* by Christian Davenport, I also consider Elon Musk one of my moonshot heroes. As a cofounder and the CEO of SpaceX, he has sparred most with the establishment to secure the right of private companies to get involved in space exploration, including suing NASA for signing no-bid contracts which illegally undermined competition. Suing them eventually made it possible for SpaceX, along with Jeff Bezos's Blue Origen, Richard Branson's Virgin Galactic (two other moonshot heroes), and other companies to reinvigorate our accelerating advancement into space.

MY MOONSHOT HEROES

Musk is also planning our species' first mission to Mars within the next decade because he believes humanity must inevitably spread to other planets if we're going to survive. This is but one example of his drive to solve our biggest challenges, including global warming, which is why he built an electric car in 2008, the start of Tesla, which has since become a top car company in the world. To compete with his success, every other car company is now planning to go 100 percent electric, signaling a looming end to the internal combustion engine. Tesla has also completed a battery Gigafactory, has others under constructions, and plans to build as many as twelve. A hundred such factories would give us enough battery power to supply the energy needs of our entire planet. All this, in addition to SpaceX now being the go-to company for delivering cargo to the International Space Station, makes Musk a moonshot hero.

The moonshot hero most inspirational to me is hardly a household name: Martine Rothblatt. If you *have* heard of her, it's probably because she's the founder of Sirius XM Radio, which she created because she wanted to connect the world, in this case, with satellite communications. In a 2019 interview with Peter Diamandis, she told the Abundance Digital Community, "People said it's impossible. It would cost hundreds of millions of dollars. It's illegal. There are laws prohibiting countries broadcasting satellite information into other countries." But she didn't let any of this dissuade her from pursuing her crazy moonshot, and today says, "It gives me immense joy now to know there are hundreds of millions of people who listen to our satellite communications signals. I've been hugged by women in remote villages who said if it wasn't for the Sirius radio station that they would go mad, they would have no access to information." She's also met kids in remote corners of India who were able to educate themselves and go to college because of Sirius's tele-education services.

You'd think creating a worldwide communications network would be enough of a moonshot for one person, but when Rothblatt's 10-year-old daughter was diagnosed with a rare, untreatable, and fatal lung disease, Primary Pulmonary Hypertension, she decided she'd find a treatment herself. She stepped down from leading Sirius XM, got a stack of biology books and articles, and began educating herself, an attorney, about a subject she knew little about. Amazingly, she did discover a drug that could successfully treat her daughter's rare condition, got it into clinical trials, approved by the FDA, and her daughter, who'd been given 2 to 5 years to live, is now well into her adulthood. Thanks to Rothblatt there are now 40,000 people living with this condition that only a short time ago would have been dead.

But treating her daughter's condition wasn't enough. Rothblatt wants to cure her by giving her a new set of lungs and has founded United Therapeutics to help make it happen. The medical company has developed technology that brings dead organs back to life and has already saved over 1,500 people with them. It also genetically edits pig organs to prevent them from being rejected by human bodies, effectively humanizing them; uses the collagen of a pig's organs as scaffolding for growing organs out of human stem cells; and uses tobacco leaves that have been genetically modified to express human collagen, then uses it to 3D print organs using no animal products whatsoever. How's that for restorative justice, printing healthy lungs out of tobacco leaves? Most of these techniques are expected to be approved and become the standard in transplant treatment in the early 2020s.

And if that's still not enough, Martine Rothblatt is also concerned that 5 percent of the world's carbon pollution comes from airplanes, so she's invented an electric aircraft, is already in the *Guinness Book of World Records* for the longest electric helicopter flight, 25 nautical miles, and plans to have a cross-country electric airplane flight in 2020. She expects the technology to be approved by the FAA and be in widespread use by 2024.

Martine Rothblatt, Elon Musk, Allan MacRae, Astro Teller, Peter Diamandis, a tattoo artist, mechanic, and a dentist, all prove what we can accomplish when we want to solve our greatest problems and grandest challenges, when we want to help others, no matter how crazy our ideas. And this is truer for those of us alive today than it's ever been. Maybe it's *only* true of us because we now have free access to whatever knowledge we need to figure things out, and to the affordable technologies necessary for carrying out our ideas. If enough people like your crazy idea, you even have access to crowdfunding, which is expected to rise to $300 billion by 2025 and has already funded hundreds of thousands of projects.

The biggest problems we face are no longer out of our control. We don't have to depend on others to solve them, not on politicians, experts, or billionaires. Poverty, racism, sexism, war, crime, illness—we can solve them all. We don't have to settle for just pointing a finger at someone to blame or looking up to those we hope can save us. We can solve these challenges for good, along with millions of others around the world who are also undertaking moonshots to do the same. Today we really are restrained only by the limits of our own imaginations. So think big. Think crazy. Think ten times, a

MY MOONSHOT HEROES

hundred times, a thousand times better than now. Be a moonshot hero. Help a billion people. Change the world.

---

[1] *Jain, Naveen. Moonshots: Creating a World of Abundance (p. 51). John August Media, LLC. Kindle Edition.*
[2] Ibid., p. 16.
[3] Diamandis, Peter H., & Kotler, Steven, *Bold: How to Go Big, Create Wealth, and Impact the World*, Simon & Schuster, New York, NY, 2015, p. 81.
[4] Ibid.
[5] 2018 A360 Archive / Categories / Module 6: Moonshots
[6] https://www.xprize.org/
[7] Diamandis & Kotler, ibid., p. 250.
[8] Ibid.
[9] Ibid., p. 253.
[10] https://www.xprize.org/
[11] MacRae, Allan, & Zehr, Howard, *The Little Book of Family Group Conferences: New Zealand Style*, Good Books, Intercourse, PA, 2004, Kindle Version, Chapter 2 (14%).
[12] Diamandis & Kotler, ibid., p. xii.

THIRTEEN

FROM RICHES TO RAGS

Once upon a time, there lived an Emperor who valued his silver and gold more than almost anything else. One day a wizard used the magic of alchemy to create a new kind of metal, shinier than silver and more precious than gold, and presented it as a gift to his imperial Lord. The wizard explained that he'd used his magic to pull the exotic metal out of ordinary clay, using a spell known only to himself. But the gift did not make the Emperor happy. He feared that if others learned of this rare new metal, all the gold and silver he'd amassed during his various war victories might become worthless. So he ordered the wizard be executed before anyone could learn his secret.

If this sounds like the gruesome end to one of the Brothers Grimm's nightmarish fairytales, it isn't. In fact, it's no fairytale at all. It's a true story, originally told by Pliny the Elder, a 1st century Roman historian. The Emperor was Tiberius, more infamously known as Julius Caesar. The wizard was his goldsmith, and the metal he extracted was aluminum, which at the time was the most abundant yet least obtainable metal on Earth. It's estimated there's twice the amount of aluminum in the Earth's crust than iron. But after the beheading of Caesar's goldsmith, the secret to its extraction remained unknown for nearly 2,000 years.

It was so rare, in fact, that only 150 years ago, another emperor and conqueror, Napoleon Bonaparte of France, had a set of aluminum utensils made that he reserved for only his most honored guests. His ordinary guests had to settle for tableware made of gold. Aluminum was once considered so precious that the French displayed aluminum bars next to their Crown Jewels, and the United States topped the Washington Monument with it in 1884. Just two years later, engineers discovered they could easily and cheaply extract the elusive metal by simply sending a direct electrical current into its ore—a process known as electrolysis. By 1888, Alcoa, the world's largest aluminum manufacturer, was producing 50 pounds a day. As a 2010 *Slate.com* series on the elements explains, "Within 20 years, it had to ship out 88,000 pounds per day to meet demand. As production soared, prices plummeted. In the mid-1800s, the first aluminum ingots on the market went for $550 per pound. Fifty years later, not even adjusting for inflation, it cost 25 cents for the same amount."[1]

Today this cheap metal still rests atop the Washington Monument, but you don't have to be an Emperor's most favored guest to eat from it. Just go to any convenience store, buy a pop, then toss the can away when you're done. It's so abundant that instead of reserving it for royal cutlery or for accompanying the Crown Jewels, we now make entire airplanes out of aluminum. The Surmet Corporation has even invented transparent aluminum that's harder than sapphire, an inch of which can stop a 50-caliber bullet better than six inches of bulletproof glass. It's still expensive to manufacture, but there are already plans to replace the International Space Station's windows with it to prevent them from scratching. In time, like aluminum cans, this lightweight material is expected to become so abundant and affordable that it will replace most uses for glass. Can you imagine scratch-proof eyeglasses? Or car windows that can't be chipped, scratched, or smashed? In the meantime, this once-rare and precious metal already has many valuable uses and has become so abundant that, "Today," as Naveen Jain says, "we throw away aluminum cans. You can't pay people to recycle them."[2]

The moral of this story is that it's possible for things once scarce to suddenly become abundant with just a little ingenuity. *Ingenuity*, by the way, means thinking about things in original ways. It means getting creative. It means not dwelling on what we don't have, but on how to have what we want. Oddly, it's often those who live with scarcity that are best at creating abundance, at making something out of nothing. Need leads to desire, desire to creativity, and creativity to abundance. "Necessity," as the old adage goes, "is the mother of invention."

Several years ago, my spouse and I tried a vegan diet, at which point some of our most favorite foods, especially desserts, became scarce because there are hardly any desserts that don't include milk, eggs, and butter. So, Peggy, the baker in our family, did some research and learned how to make some of the moistest, richest, most delicious cakes I've ever had. Vegan desserts remain a favorite in our home because they often taste better than those loaded with animal products.

What may surprise you is that just a couple of generations ago some of our grandparents were making the same kinds of desserts, not because they wanted to avoid dairy and eggs, but because these foods became scarce during the

Great Depression and World War II. They didn't call them "vegan cakes" back then, but gave them more sardonic names, like "Depression cakes," "War cakes," or simply, "poor man's cake." Sugar was also scarce in those days, so they learned to sweeten things with syrups or boiled raisins. The point is, instead of giving up and saying, "No more desserts: the ingredients are too expensive and in short supply," they got creative and found a way to have their cake and eat it too.

There's a large pyracantha bush in my neighborhood that remains covered with clusters of small red berries through most the winter. But midway through the season, when other food becomes scarce, the deer resort to eating them. By the time they're done, the bottom half of the bush is green, but the top half, which they can't reach, remains red. It's as if a straight line has been drawn around the circumference of the bush. In a time of scarcity, deer are a species that must literally settle for the low-hanging fruit, and by the time it's gone they can only hope spring isn't far behind. But our species doesn't have to settle for low-hanging fruit. We have the smarts to make stepladders, construct cherry pickers, and build rockets that can carry us into the heavens where there are more stars than we can possibly grasp.

These examples help us differentiate between two mindsets, one of scarcity and one of abundance. One mindset says, "There's not enough and that's all there is to it." The other says "There's not enough, let's solve the problem." The latter doesn't mean living beyond our means. It doesn't mean using more resources than we need or are available. But it doesn't mean settling for the way things are either. This is why my least favorite saying attributed to Jesus is, "The poor will always be with you."[3] Over the years I've heard many people use this verse to justify doing nothing about poverty because it makes poverty sound like it's divinely ordained.

But why should we settle for poverty or take it for granted the poor are just a part of life and there's nothing we can do about it? I much prefer the Jesus who talked about having enough for today, "our daily bread," but did so with a mindset of abundance, not scarcity. "Don't worry about packing your coffers with more than you need because there's always more than enough to go around. Look at the birds who have way more than they can eat, and at the flowers that are clothed more splendidly than the wealthiest of kings." There can be more than enough to go around for everyone if we think creatively. "I came that they may have life, and that they may have it abundantly,"[4] Jesus is also reported to have said.

My own mindset on this was challenged recently while reading Naveen Jain's book, *Moonshots: Creating a World of Abundance*. Jain writes that abundance "is different from the idea of sustainability. We have to create more of what we need rather than consume less of what we have. Sustainability is a synonym for conservation of scarce resources, and you cannot achieve sustainability with conservation alone."[5] At first, as an environmental activist, I was bothered by this notion because it shook my paradigm. I can still remember the time, many years ago, hearing theologian Matthew Fox say, "There's a new word for justice—*sustainability*," even before most anyone else was using the term. Since justice is synonymous with balance and harmony, I've considered sustainability a kind of justice toward the Earth ever since. But I don't think sustainability is simply about living within our means. It means not taking more than we need, not more than our daily bread, especially not at the expense of others or the environment.

In this sense, sustainability and abundance can work together. There's no sense making pigs of ourselves when we have more than enough, but there's also no sense settling for poverty, hunger, homelessness, unclean drinking water, polluted oceans, and dirty air when we can create ways to make sure everyone has enough of everything they need to thrive. But that's exactly what the scarcity mindset settles for because it assumes there's seldom enough to go around. Resources must be rationed. Some have more than enough while others must go without. An abundance mindset figures out a way to solve the underlying problem of scarcity to begin with.

Realizing, for example, the age of cheap oil energy is ending, Abu Dhabi in the United Arab Emirates, one of the oil-richest nations on the planet, is building the world's first entirely green city. *Masdar*, an Arabic word for "source," will be completely solar, leave no carbon footprint, consume no fossil fuels, be bicycle and pedestrian friendly, and will be home to 50,000 people with another 40,000 employed there.

Jay Witherspoon, the technical director for the $20-billion project, now scheduled to be completed in the early 2020s, says it's based on the principle of OPL—One Planet Living. One Planet Living, a global initiative begun by BioRegional Development and the World Wildlife Fund, recognizes we're currently using 30 percent more resources than Earth can replace. In *Abundance*, Diamandis and Kotler say OPL means, "If everyone on this planet wanted to live with the lifestyle of the average European, we would need three planets worth of resources to pull it off ... If everyone on this planet wanted to live the lifestyle of the average North American, then we'd need five planets to pull it off."[6] Even now there's not enough to go around. We can conserve resources, and should, but this will only bring the quality of life down for some, not up for most.

This is why an Abundance mindset is key to solving our greatest challenges. We've been using up a limited supply of dirty coal and oil for 200 years. It's made some people and companies filthy rich, some countries energy rich, life more convenient for billions of us, but has been ruining the planet for all of us in the process. Even our preindustrial, prescientific ancestors knew it's really the Sun that gives both life and energy. Yet today, even in our industrialized scientific age, 90 percent of our total energy use, about 16 trillion watts a year, comes from non-renewable energy. But there's 5,000 times that much solar energy annually reaching the Earth's surface. As *Abundance* says, "it's not an issue of scarcity, it's an issue of accessibility."[7] So what if, instead of strictly reducing our energy consumption, we find a way to better utilize a tiny fraction of the energy our life-giving Sun bathes us in every day, making affordable energy as available as aluminum foil?

The same mindset can be applied to clean drinking water. With few exceptions, like Flint, Michigan, most of us living in North America don't worry much about the water we drink. But according to the United Nations, more than 2 billion people around the world lack access to safe water; each year unsafe water kills 340,000 children under age 5; and water scarcity impacts 4 out of 10 people every day. Meanwhile, 70 percent of our water resources go into agriculture, 75 percent of the water used in industry is used for energy production, and 80 percent of polluted wastewater goes right back into the ecosystem.[8] All of this represents a serious problem given that potable drinking water is one of our planet's scarcest resources. Although ours is a mostly water planet, 97 percent of it is undrinkable saltwater. Of what's left, 2 percent is frozen in glaciers and ice caps or is too deep underground to reach. That leaves only 1 percent of the water on Earth that we can drink, if we're lucky enough to access it and can guarantee it's safe.

A scarcity mindset says conserve what we can, purify what we can for those countries that can afford it, but the rest must go without, which is today's status quo. But what if, instead, we start looking at the problem through the lens of abundance? Instead of looking at water as a scarce resource, let's recognize that it's the most abundant resource on Earth. Hell, it's the most abundant resource in the Universe! Its first and most abundant element is hydrogen, from the Greek word meaning, "water maker." Let's learn the lesson of aluminum by continuing to improve our methods of desalination so we can drink water from our overflowing oceans, as well as providing plenty of clean, green energy to industry, while purifying the fresh water on our planet in the process. We don't have to settle for the low-hanging fruit here. We can climb Mount Everest. We can reach for the stars.

The mindset of Abundance can also be applied to how each of us lives our individual lives. We can look at life through the lens of scarcity or through the lens of abundance. Some might say that's my North American privilege talking, but I grew up living in poverty and lived in poverty much of my life. Still, I never knew it as a child because scarcity itself led me to think creatively and innovatively. I'd pull bent rusty nails from old boards, straighten them, then use both the nails and the boards to build my forts, or to make the best go-carts a kid could wish for.

I remember my mother, who tried to compensate for our lack by baking sweets for my siblings and me, sometimes didn't have enough in the house to even do that much. Once in a while, when she was desperate, she'd make my favorite treat of all. She'd mix flour and water together with a little lard to make a basic dough, roll it out, sprinkle it with a little cinnamon, roll it up, slice it into small pieces, bake them, then cover them with whatever sugary glaze she could come up with. I loved them! I used to beg her to make them. But she seldom would and wouldn't explain why. When I got older, I realized it was because she was embarrassed. Those treats reminded her of how poor we were. But for me they were the stuff of fortresses, go-carts, and all the other riches of my childhood.

---

[1] http://www.slate.com/articles/health_and_science/elements/features/2010/blogging_the_periodic_table/aluminum_it_used_to_be_more_precious_than_gold.html
[2] Jain, Naveen. Moonshots: Creating a World of Abundance (pp. 108-109). John August Media, LLC. Kindle Edition.
[3] Matthew 26:11
[4] John 10:10
[5] Jain, ibid.
[6] Diamandis, Peter H. & Kotler, Steven, *Abundance: The Future is Better than You Think*, Free Press, A Division of Simon & Schuster, New York, NY, 2012, p. 5.
[7] Ibid., p. 6.
[8] https://www.un.org/en/sections/issues-depth/water/

FOURTEEN

THE GOOD NEWS

When I consider our species' awkward form, I wonder how we've lasted so long and done so well. We have oversized heads atop an unstable pair of stilts, with a weak body and spindly arms between them. We're not fast enough to outrun predators or catch up with prey. We can't fly away or scurry up trees toward safety, nor display sharp teeth and claws to defend ourselves. Yet we've not only managed to survive, we've become the dominant species on our planet.

Bipedal animals, in general, aren't well-suited for survival. Most of them, like some of the dinosaurs and a few amphibian species, have already gone extinct. All birds are bipedal but, with only few exceptions, flight, not walking, is their primary method of locomotion. Just walking didn't work out so well for the dodo. Among mammals, bipedalism is mostly a feature of us primates, but even nonhuman primates mostly walk on all fours. Monkeys walk like cats, and other apes on their knuckles. Other kinds of mammals, like racoons, rats, and meerkats, demonstrate limited bipedalism, but these only stand on their hind legs long enough to look out for potential threats before hunching down again to move forward on all fours.

Only humans stand completely upright when we're on the move, a posture of fear and vulnerability in other animals. Even a rattlesnake becomes as upright as possible when threatened. We embody what fear looks like in nature, moving about as if we're always on the lookout for danger, which I suspect has much to do with our tremendous success. In the past, when humans were vulnerable hunters and gatherers, exposing ourselves to danger by entering into open plains and fields from which we could not outrun a hungry predator, we had to remain hypervigilant, always on the lookout for the slightest threat.

Fear also has a special place in our heads: the amygdala, two almond-shaped clusters of neurons connected to the hippocampus, the emotional part of the brain, that are responsible for fear, anxiety, and defensiveness. So humans both embody what fear looks like and are wired to look at the world through a mindset of fear. We've evolved to look at what's wrong with the world, not at what's going well. Even though we don't have to worry about nearly as much imminent danger as our ancient ancestors once did, our bodies and brains are still designed to cope with a frightening and dangerous world. We are natural-born pessimists, wired to look for the worst.

Only a few years ago, in 2007, psychologist Kevin J. Flannelly and his team of researchers introduced a particular branch of Evolutionary Psychology called *Evolutionary Threat Assessment Systems Theory* (ETAS) which suggests the neurological structures that help us evaluate the potential threat in any situation are the initial motivating factor in all we think and do. "A central feature of psychiatric disorders," Flannelly explains, "is a primitive concern about one's own safety and the dangerousness of the world, with different psychiatric disorders representing the response of threat assessment systems in the brain to different kinds of potential threats."[1] Cognitive therapist Paul Gilbert, a proponent of ETAS theory, says "the most important question faced every day by all animals, including humans, is whether their immediate environment is dangerous or not."[2]

As a result, we tend to mistrust the rare optimists among us and consider them naïve and out of touch with reality. But if we have evolved to naturally look for and expect the worst, maybe it's our pessimism that isn't always based on reality. Maybe we don't often recognize the good things happening around us simply because we're not wired to see them, finding them hard to believe even when they're pointed out.

Even so, there have always been optimists among us, those whose disposition is to look eagerly toward the discovery of better fields, plains, and futures where we don't have to worry about crouching tigers and hidden dragons. More than 500 years ago, Sir Thomas More wrote a book about the political systems of an imaginary island nation he called *Utopia*. In the recent book, *The Fourth Age*, Byron Reese reminds us *utopia* is a word that means, "no place." Yet even in his dreary 16th century world, there were others like More who weren't afraid to look forward to better possibilities.

> When you read some of oldest of the utopian literature from the sixteenth and seventeenth centuries, you can only imagine how outlandish the worlds that these authors envisioned must have seemed at the time. Sir Thomas More's 1516 book *Utopia*, which gave us the word 'utopia,'

describes a land with religious freedom for all. *Civitas Solis*, written in 1602, imagines a place with no legal slavery, while *Adventures of Telemachus* of 1699 describes a utopia with a constitutional government.[3]

At the time works like these were written, what they envisioned were merely pipe dreams that existed "no place." Yet today, probably because someone dared imagine such impossibilities to begin with, we have separation of Church and State, slavery is a violation of international law, and most countries are constitutional democracies. Since the end of World War II, when there were only 10 democracies on Earth, democracy has gone 10X. That's right, there are now nearly 100 democracies round the globe. According to the most recent Pew Research findings, "As of the end of 2016, 97 out of 167 countries (58%) with populations of at least 500,000 were democracies, and only 21 (13%) were autocracies."[4] An idea that was once thought so crazy as to exist "no place," save the naïve imaginations of out-of-touch dreamers and philosophers, has become the dominant form of government on Earth. As Reese explains:

> In the nineteenth century, more utopian literature was published that contained outlandish ideas such as universal education, the legal equality of men and women,

governmental safety nets, and preventative health care. Those ideas are not regarded as crazy anymore, and our modern world is on a trajectory to achieve them. And when we achieve them, we will have come up with still more crazy ideas, until one day we will wake up and find that we cannot imagine a world any better than the one we live in.[5]

Optimism isn't the result of naïve speculation, or even dreaming impossible dreams. It occurs when we manage to overcome our biological tendency toward fear. When this happens, we're more able to look at the facts and see the world as it exists beyond our innate instincts and mindset, beyond fear. "The fact is that by every objective measure, the world is getting better," Naveen Jain writes. "Much better."

> The number of people living in poverty has never been lower. There are far more democracies in the world. Literacy has reached an all-time high. Higher food production and lower costs have put a massive dent in world hunger. Infant mortality rates have plummeted. We're on an accelerated path to electric cars and far less consumption of fossil fuels. Sanitation standards, life expectancy, air quality—it's all improving. Heck, even the giant panda is no longer endangered.[6]

My own innate pessimism began changing several years ago when I was asked to chair a committee to create a Restorative Justice pilot program in Louisville, Kentucky, on behalf of the County's Disproportionate Minority Confinement Advisory Board. At the time, I didn't hold much hope for success because I knew we were going up against the legal establishment, which I imagined was carefully guarded by people who didn't want to relinquish their power and position, regardless of how many continued getting hurt in the process. Back then I still believed most people are self-centered, a pessimistic view of human nature, and I expected the worst from them.

But in a very short time my opinion began to change. In fact, I never tell lawyer jokes anymore because the many attorneys I encountered during this period helped redeem my faith in the innate goodness of all humanity—if being good means caring for the welfare of others, which is the core of the humanistic ethic. Each Wednesday morning, prior to the start of court, our committee met at the courthouse at 8 a.m., in the judges' conference room, to work on our daunting project. The committee included the County Attorney, the Public Defender, sitting and retired Judges, the Deputy Police Chief, and several attorneys, all of whom regularly expressed their desire to fix a broken system that ruins the lives of people on both sides of the bars. They were among the most compassionate people I've ever met, who often found themselves trapped in an overwhelming system of injustice. When the State government found out what we were up to, they were so excited that, despite the economic downturn at the time, they voluntarily offered to help fund whatever program we developed. Today, because of their compassion, Louisville has a successful Restorative Justice program in place keeping teenagers out of the school-to-prison pipeline.

After this experience I began wondering about others I may have misjudged. I know politicians get a bad rap, but is it possible most of them, just by virtue of being human beings, enter into the field because they genuinely want to make a positive difference in the world by helping others? Doing so is seldom easy, however, given the divergent interests and ideologies out there. Sometimes compromises have to be made to get anything accomplished, which usually doesn't make anyone look good. I wonder if it's just the nature of the beast, that politicians can't make any decision that pleases everyone, so they always end up being despised by someone. This isn't to say there aren't some underdeveloped human beings who are drawn to politics because they haven't outgrown their childish authoritarian stage of development; but, as a humanist who rejects the notion of human depravity,

I have to conclude most politicians are genuinely altruistic people.

I also realize corporations are not persons and should not enjoy the same rights as those who are. Yet they are begun by persons, overseen by persons, employed by persons, and can give purpose, creativity, resources, and opportunity to the lives of persons. All millionaires and billionaires, on the other hand, are persons, though most persons are not millionaires and billionaires. Is it likely, just by virtue of being persons and being run by persons, that most billionaires and corporations care about the welfare of all persons?

By the 1990s many University business schools were teaching the relatively new field of Business Ethics. It was as recently as 2015 that I began noticing the difference a generation of teaching ethics to budding entrepreneurs is making. After hearing, within a short period, of a bank voluntarily raising its minimum wage to $15 an hour and calling on the entire banking industry to do the same, and of another company's CEO reducing his $1.1 million salary to help raise his company's minimum wage to $70k per year, and of several major corporations, including Apple, Subaru, Gap, Angie's List, and the NBA, among others, boycotting the State of Indiana after it passed a law making it legal for businesses there to discriminate against gays and lesbians, my interest peaked. Instead of looking for the worst in corporations, I began looking for the best.

During a 2019 conference Peter Diamandis told a group of entrepreneurs, "It pisses me off to see all these billionaires out there who are taking their cash and hoarding it instead of using it to go and transform the planet. There is no problem we cannot solve. Within the next 10 to 20 years we have the potential to give every man, woman, and child on this planet access to all the education, healthcare, energy, food, water they want." He then assures the pessimists, "You might be barraged by the negativity of the news network; put that aside and look at the data and realize how powerful you are":

> What is it that you want to do that at the end of the day for your legacy that makes your kids and the world proud of you? It brings joy and it brings challenges. An unchallenged life is not worth living. If you find your life easy, you're on a beach in Hawaii, amazing. Come back. Come back, the world needs you ... This is the world we're in guys. It's not the government. It's not the robber barons. It's us. It's you. It's me. Any one of us can take on and solve a massive challenge ... Do not dream small dreams. That's not what we're here to do.[7]

If you think Diamandis is a dreamer, remember how crazy utopians like Sir Thomas More sounded in the 16th century when talking about some of the realities we now take for granted. What if I were to say I see a future in which there's free renewable energy for everyone, in which we've reversed the worst impacts of global warming, cleaned the plastic out of our oceans, and human population growth is declining, not exploding? Would you call me a naïve optimist? A wishful thinker? Would you think me mad?

Then you might be surprised to learn inventor, writer, and energy expert, Ramez Naam recently noted that only 10 years ago it cost 10 to 12 cents to produce a kilowatt hour of solar energy, twice the cost of coal and gas energy. Today it's just 2 cents and is quickly moving toward a penny per kilowatt hour. Naam says, "We are nearing the point where we are not building any new fossil fuel power plants. Basically, we're there on coal with a few exceptions in Southeast Asia."[8] Between 2014 and 2019, 75 percent of coal companies around the world have gone bankrupt and, with increasing efficiency and plummeting prices of batteries, it's now cheaper to build electric cars than gas powered, which is why every major car manufacturer is moving in that direction. "So, we're looking at electric vehicles being cheaper on upfront costs within about five years and most importantly we think the peak of internal combustion engine sales is on the horizon."[9]

How does it make you feel to hear someone say the era of fossil fuel use is quickly ending, that coal is over, petroleum use is on its way out, and renewable energy will soon be almost free for everyone in the world? Dubious? Difficult to believe, even with the numbers right in front of us?

Let's turn to the human population explosion, another major factor in the destruction of our environment. We've all heard it reached 7.3 billion at the start of this century and has been predicted to be 11.2 billion by its end. But what if this is no longer true? What if, as mentioned earlier, it's difficult to imagine we'll even hit 9 billion? Right now, the global birth rate is near to or lower than the necessary replacement rate in most places, including in India, North Africa, and Southeast Asia. The only reason our population continues to grow is because the average human life expectancy is increasing.

According to the recent book on the subject, *Empty Planet: The Shock of Global Population Decline*, this is largely due to the urbanization of the human population, which has led to the education of women and widespread use of contraception. As its authors say, "Urbanization and the empowerment of women are having the same effect on developing countries that they had on developed countries, only everything is happening much, much faster. Across the planet, birth rates are plunging. That plunge is everything. That plunge is why the UN forecasts are wrong. That plunge is why the world is going to start getting smaller, much sooner than most people think."[10]

Even hearing this, it's hard to believe it's true. We want to believe it, but we're afraid to. Yet if we can see past the fear and look more objectively at the world, we might envision a less frightening future. "What will the world be like for a child born today when she reaches middle age in a time of population decline?" *Empty Planet's* authors ask. "We believe there will be much about that world to admire. It will be cleaner, safer, quieter. The oceans will start to heal and the atmosphere cool—or at least stop heating."[11]

Speaking of the oceans, have you heard about the plastic-eating bacteria scientists discovered in 2016? The bacteria, found in a waste dump in Japan, has evolved to eat plastic. While testing it to figure out how it manages to produce an enzyme that degrades it, scientists unintentionally made it faster. Plastic normally takes several hundred to about a thousand years to degrade, but this enzyme can break it down in just a few days. As one researcher at the U.S. Department of Energy's National Renewable Energy Laboratory says, "Engineered enzymes that break PET down to its building blocks would enable the ability to do full bottle-to-bottle recycling,"[12] which will also decrease the need to use new oil for plastic production.

There's so much else we have to worry about today that leads us to imagine frightening futures. Instead of tigers and vipers in the fields before us, we face global warming, population explosion, nuclear war, terrorism, and so many other reasons to worry about venturing forward. But if we can see past our fears, maybe we can also envision a world without such worries. Can you imagine a world without war? Or one without poverty? Or one without racism? How about one world, beyond nations, beyond nationality, in which we all recognize we are part of one human family? How about a world without disease? Or a world where people don't have to work for money because we don't use money anymore? A world in which we have everything we need as a basic right and get to choose what we do with our time and energy. This might sound like just a utopian dream, but as history has shown, yesterday's utopian dreams can quickly become today's realities.

---

[1] Flannelly, Kevin J., and Galek, Kathleen, "Religion, Evolution, and Mental Health: Attachment Theory and ETAS Theory," *Journal of*

Religion and Health (2010) 49-337-350, Published online, March 17, 2009, Springer Science+Business Media, LLC, 2009, p. 344.
[2] Ibid., p. 340.
[3] Reese, Byron. *The Fourth Age: Smart Robots, Conscious Computers, and the Future of Humanity*. Atria Books. Kindle Edition.
[4] https://www.pewresearch.org/fact-tank/2017/12/06/despite-concerns-about-global-democracy-nearly-six-in-ten-countries-are-now-democratic/
[5] Reese, ibid.
[6] Jain, Naveen. *Moonshots: Creating a World of Abundance* (p. 149). John August Media, LLC. Kindle Edition.
[7] A360 2019
[8] Ibid.
[9] Ibid.
[10] Bricker, Darrell, *Empty Planet*, Crown/Archetype. Kindle Edition, 2019, p. 29.
[11] Ibid., p. 225.
[12] https://www.popsci.com/bacteria-enzyme-plastic-waste

FIFTEEN

FACING VIRTUAL REALITY

In 2018 I discovered the wonder of virtual reality technology, which has been mind-blowing fun. But I've more recently realized that today's technology is just the latest upgrade in something that's been around a very long time. When I was a kid, a VR headset was called a View-Master, those handheld toys kids inserted a thin cardboard disk into containing three-dimensional stereoscopic photographs. While peering through its lenses, users were transported into naturally illuminated scenes. Each disc contained fourteen images, which could be easily switched by pulling a lever on its side. The most popular View-Master discs were those transporting users to famous tourist attractions around the world, places most would otherwise never see.

The View-Master was introduced way back in 1939. But some would argue VR goes back much further than this. In his book, *Defying Reality*, David Ewalt talks about recognizing VR in how the 17,000 year old paintings are positioned on the walls of the Lascaux Cave in Southern France. It's "a particularly fitting way to visit," he says, "because Lascaux may be the oldest example in human history of an attempt to create a virtual world."[1] You might not notice this just looking at pictures of its paintings, but you definitely can from inside the cave. "Just like VR headsets block the real world from view, the caverns separate a visitor from the forest above. Instead of drawing with pixels on a screen, the cave artists used pigments on rock walls."

> Lascaux's creators used the topography of the caverns to create immersion: a chamber decorated with sketches of wild horses isn't just a jumble of drawings, but a herd that surrounds the viewer. They used perspective tricks to make the illustration seem three-dimensional: the body of an ox is presented in profile, but its head is turned to face the viewer. And they used the shape of the rock to give their art depth and form; for example, a twist in a wall makes a deer appear to turn away as it dashes around the corner… Maybe

the caves were immersive entertainment, an attempt to tell a story in the most realistic way possible, to convey the excitement of hunting without the risk of getting gored or trampled.[2]

Whether with ancient paintings on cave walls or modern art hanging on gallery walls, from Shakespeare's plays to novels that excite the mind's-eye, or photographs, TV shows, movie theaters, and today's headsets, it has always been part of human nature to simulate real world experiences, as least if Ewalt is correct in saying, "all art strives to create virtual reality, to transport an audience and immerse them in a subject or story."[3]

Developmental psychologist Robert Kegan says, "what a human organism organizes is meaning," similar to what Holocaust survivor Victor Frankl famously meant when he said our "search for meaning is the primary motivation in [our] life." This is why sociologist Peter Berger further recognizes we all have an "innate need for meaning"[4] and that we "cannot accept meaninglessness."[5] It's also why we find meaning in the constellations, or by considering the meaning of our dreams, or contemplating the images on Tarot Cards and inkblots, or see Virgin Mary in the clouds and Jesus on a slice of toast. Doing so can help lead us toward better sanity, which is what Frankl's *logotherapy* accomplishes by helping patients find meaning in their suffering and trauma, and why Sigmund Freud analyzed their dreams, and Carl Jung used Tarot cards, Runestones, and the I Ching. Finding order within the chaos, creation within the static, sense out of senselessness, helps holds us steady, gives us purpose, reveals the hidden path before us.

    We make meaning by creating virtual reality inasmuch as we interpret the few bits of data we experience in a certain way. We see the world through the bias of our beliefs, the whims of our emotions, and the limitations of our senses. For these reasons, what we believe about the world is always inadequate, at best. Our feelings are subjective interpretations, not absolute or objective. Even our physical senses allow us to see and hear only a small bandwidth within a much larger range of wavelengths, and even then, only those within our immediate vicinity. No wonder we want technologies that help us see more than the little bit around us, or take us places we might otherwise never go, or show what things were like in the past, what they can be like in the future, and what they are like right now on Mars, or inside a black hole, or in some other galaxy far far away.

    But sometimes our natural ability to make meaning also gets us into trouble by interpreting reality in ways that are baseless and misguided. This can lead us to embrace paranoid conspiracy theories, or cling to old patterns of belief that have been disproven and are harming us, others, and our world. The virtual realities we construct for ourselves, like race, nationality, and gender, can be delusional: delusional thinking we often attempt to force upon others, forbidding the expression of their own realities just so we can feel more secure in our own.

In this sense Fox News is virtual reality. So is MSNBC. Rush Limbaugh is a professional delusionist. But in a sense, almost all news media is delusional, not because it's all false, but because it's too narrow. It tunes in only a small bandwidth of reality that we too easily mistake as the whole of reality; and a partial truth can be as misleading as an outright lie. In general, the media focuses on negative news, the worst things happening, the most frightening and troubling stories going on. The U.S. news media is its national amygdala: because it's always looking out for what's wrong. And when we mistake its narrow channel of negativity and fear, through which all its information flows, as the whole of reality, it's easy to believe we are living in a frightening and dangerous world. So fear becomes our matrix too, the View-Master, the headset, through which we see the world: a virtual reality that leaves us blind to all the goodness and good people also in the world.

Every now and then I hear someone express concern that soon we'll all be spending more time in virtual reality than in reality. But we are already spending most of our time in virtual reality. Reality is at once the most abundant resource in the Universe and its rarest gem. Indeed, it is the only resource in the Universe. For whatever *is* must be real, even those realities we cannot yet explain. But reality is also impossible for us to fully comprehend. At best, like Moses tucked within the cleft of the rock to catch but a glimpse of Yahweh's backside, we use science and reason to infer what's real from the evidence it leaves behind. We can't physically observe quantum particles, but we can use accelerators to smash them together and examine the damage they etch into small plates. We don't live long enough to observe major geological epochs, but we can observe the different sedimentary layers time presses into rocks and cliffsides. We can't watch animals evolve into new species, but we can learn how they came to be by looking at fossils. If we're genuinely eager to learn the truth, to face reality, we might catch glimpses of it, but can never fully comprehend it.

The problem is not that we will soon be spending too much time in VR, but that we already are and most of us don't realize it. The real issue is that we think virtual reality is something new when it's all our species has ever known. The wish to impose one's own reality upon others, because it's mistaken as the whole of reality, has led to many of history's worst injustices. This is what the Crusades, Inquisitions, heresy trials, witch burnings, McCarthyism, and the Cold War were all about: one VR program trying to overwrite another. It's also what today's culture wars are all about, liberals and conservatives each trying to control the narrative of others so they can control the stories we tell ourselves about what is and isn't real.

There's also the psychological difficulty most of us have dealing with *cognitive dissonance*, which is a lot like putting on a VR headset and becoming blind to everything outside our limited virtual experience. As science writer Jonah Lehrer says, "The default state of the brain is indecisive disagreement, various mental parts are constantly insisting that the other parts are wrong."[6] Neurologist Robert Burton says our beliefs aren't determined as much by reason as they are by a kind of inner-committee "representing all your biological tendencies and past experiences."[7] We don't usually realize this because it "meets behind closed doors, operating outside of consciousness in the hidden layer"[8] of our minds. Instead of dealing with the noise of endless possibilities within us, we adjust the dial until we hear only one channel, one truth, one reality, becoming unable to entertain any others. "We all silence the cognitive dissonance through self-imposed ignorance,"[9] Lehrer says.

At this point, I don't think VR headsets are going to cause us to be any more out of touch with reality than we already are. On the contrary, I suspect in the next few years VR is finally going to succeed in accomplishing something we've been striving for since the underground theatre of Lascaux was designed 17,000 years ago—truly transporting us to places beyond the limitations of our local realities. VR will soon make philosophers out of us all by forcing us to question the very nature of reality, freeing us from the small worlds we've fashioned for ourselves or that others have fashioned for us.

If the authorities catch wind of it, they'll probably give up fighting illicit drugs and start a new War on Headsets. Dominators don't like having their paradigms questioned, especially by alternate realities that help better unify the world. Yes, losing our grip on reality, or at least on what we have accepted as reality, isn't always fun, but whether we are prepared for it or not, our reality is about to be disrupted in unprecedented ways. The status quo is about to become the *static* quo.

Today's VR technology already allows us to scuba dive, fly jet airplanes, relax by campfire beneath the Northern

Lights, watch sports from the floor or field of a live game, go bowling, travel the solar system, roam about the surface of Mars, or fight to survive the zombie apocalypse in the comfort of our own homes. Yet it has only been available to consumers since 2015. More amazing is that it also allows us to meet and form friendships with real people from around the world by hanging out with them in virtual spaces. There's even a Real Estate company based in Bellingham, Washington that has more than 600 employees and no physical buildings. It has virtual offices in a virtual town, and even has a virtual park and soccer field.

Augmented reality isn't as far along as VR yet, nor as affordable, but it soon will be, and when it is, it may be more revolutionary than virtual reality. Augmented reality blends with normal reality, allowing us to have access to people and things that appear to be with us, but materially are not. Instead of calling someone on our physical phone, for example, we can look at our hand and have a virtual phone appear. The person we're calling can appear right in front of us. Coupled with the new 5G internet speed, the VR and AR graphic quality will appear to be as sharp as normal reality, and the virtual environments we create can be enormous, allowing us to move around in them for hours. Imagine a wall of your home turning into a window to anywhere in the world. You can look out over New York City in real time, or observe a tropical beach on another continent, all while experiencing the same photons there hitting your retinas wherever you are.

If wearing headsets or glasses seems cumbersome, a company called OTOY has successfully developed 2 by 2 square-foot holographic panels that can cover a wall, creating a virtual window to the world, or cover an entire room, floor to ceiling, to create a grid that looks like the holodeck on *Star Trek*, that you can walk about in wearing nothing—headset, glasses, or otherwise. This technology is currently priced out of sight for most people, but it does exist, and it's only a matter of time before the price drops and it becomes commonplace.

What excites me about all of this, though it probably terrifies some others, is what we experience in VR and AR will expand our understanding of reality. There's a difference between the world as we perceive it and the world as it is. Our human senses are like that radio tuner, allowing us to perceive a very narrow range of a much larger spectrum. This is why Descartes famously wondered if the world outside our heads might not even exist— if it isn't a dream or the trick of some evil demon—which led him to conclude the only thing he could be sure of was his own existence, *cogito ergo sum*: "I think, therefore, I am." Philosopher Bertrand Russell felt more certain there is a world that exists outside ourselves, but understood we are unable to truly perceive it. What we perceive, rather, is what he called "sense data," something we come into contact with that our minds then interpret into images we can comprehend. X-ray machines, MRIs, ultrasounds, infrared and ultraviolet lights, radar detectors, metal detectors, microscopes, and telescopes are just a few of the technologies we have that remind us there is much more to the world than meets the naked eye.

Openwater, for example, is a company that's developing wearable technology to replace multi-million-dollar MRI scanners, which currently puts the technology out of reach for most of us and isn't usually covered by insurance until it's already too late. Dr. Mary Lou Jepson, the company's Founder and CEO, says it will be 1,000 times cheaper, a million times smaller, and have a billion times better resolution than today's MRIs. Combining the technologies of infrared light, holography, and ultrasound, it will allow us to wear clothes that interface with our computers to constantly monitor what's going on inside our bodies and catch the first signs of something going awry before it becomes a big deal. For our purposes, its resolution is so high that, combined with AI, it can also be used to recreate the images in our heads, not only those we see with our eyes, but also those we dream about or simply imagine, because they all occur in the same part of the brain. In other words, as far as our brains are concerned, there's no distinction between what we see with our eyes and what we imagine. Our brains are interpreting sense data to create a virtual reality we can perceive.

Plato had it right—we sit in a cave watching shadows on the walls that we believe are all there is to the world. It's not much different than the virtual bubbles today's VR software create for us. They can provide the illusion of looking down at the entire Earth while floating outside the International Space Station, but, in reality, they're too small to hold enough information to allow much mobility. To move about very far in VR, you have to sequentially enter different bubbles. Likewise, if what we take for granted is the real and only world, some of us become extremely agitated and threatened when others question the extended realities painted on our ancient walls. We consider them insane, dangerous, or worse, because we dread the possibility ours might not be the only reality, or at least not all there is to it.

Philosophers and mystics have understood the evasive nature of reality and truth for centuries. As Taoism says, "Looked for, it can't be seen; listened for, it can't be heard; reached for, it can't be grasped."[10] Philosophers and mystics have often been treated like quacks, at best, and silenced, banished, or executed, at worst. Today's VR technology is about to change this because it will cause all of us to question the meaning of reality and the nature of truth. When we can no longer discern the difference between what's right in front of us and what only appears to be right in front of us, understanding that both are perceived by the same areas of the brain and both consist of quantum particles, it will be difficult to argue either isn't real, or that either is.

When this happens, my hope is we will all, as a species, stop being so rigid about our beliefs, and stop forcing them on others, and that those who do will no longer have power over any of us. I hope the propaganda machines erected to control the narrative and minds of others, like Fox News, political correctness, and social media, or charismatic authoritarians who influence the minds of millions with images of fear and hate, all become meaningless because they no longer make sense, because we see them for what they are, small bubbles of concocted realities too small to be true.

But we don't have to own a VR headset for this to happen. Our own brains have been fashioning virtual realities for thousands of years. It's what we do. All we need is to realize our creative minds are always at play with the world. This isn't to say reality isn't objective, that it's whatever we want it to be. That kind of thinking creates another set of problems. It only means that our interpretation of reality is subjective, and we should, therefore, remain humble about what we think we know, and respectful of those who see things differently. Like most everything else in our lives, virtual reality is evolving exponentially, which means we're getting far better at it than ever before, but it's still nothing new. It's human. It's what we do.

It is my hope that as we come to live in a world of dematerialized stuff and gain the power to easily transcend the limitations of being stuck in one location, we'll use this technology the way our ancestors in the caves of Lascaux intended: to transcend the confines of our own bubbles, to realize there's a world we never imagined possible, to transport ourselves to places just as special as our own, and to make friends in places as far away as Australia, or Afghanistan, or Southeast Asia, or the Sudan might be, and feel about them as we do our next door neighbors because our bubble of reality has become big enough to embrace the entire world.

---

[1] Ewalt, David M., *Defying Reality: The Inside Story of the Virtual reality Revolution*, Blue Rider Press, New York, NY, 2018, p. 16.
[2] Ibid., p. 16 & 17.
[3] Ibid., p. 19.
[4] Scott, Mark S. M., *Journey Back to God: Origen on the Problem of Evil*, American Academy of Religion (AAR) and Oxford University Press, Inc., New York, NY, 2012, (Kindle Version), loc. 392.

[5] Ibid., loc. 411.
[6] Lehrer, Jonah, *How We Decide*, Mariner Books edition, New York, NY, 2010, p. 210.
[7] Burton, Robert, *On Being Certain*, St. Martin's Press, New York, NY, 2008, p. 48.
[8] Ibid.
[9] Lehrer, ibid., p. 207.
[10] Mitchell, Stephen, *Tao Te Ching*, Harper & Row Publishers, New York, NY, 1988, #14.

SIXTEEN

NOTHING BUT TAXES

There's an old saying, "Nothing is certain but death and taxes." But there's a growing number of scientists who no longer accept the inevitability of the latter. This may seem strange, given nobody has ever lived much past 100 years, and most not that long. So how can anyone seriously think death is not inevitable?

Our anthropocentric view of life can be partly to blame for such certainty. All humans ever to have lived have died, which is a pretty sound reason to conclude, "all humans are mortal," an inference that famously led Aristotle to certify the death of Socrates. But is it also true that *all* living beings must die? We know there are species of flatworms, jellyfish, lobsters, turtles, and other creatures that are not biologically mortal, meaning they don't die of old age. According to a 2006 *New York Times* science article, Christopher Raxworthy, a herpetologist with the American Museum of Natural History, says "the lungs, livers and kidneys of a centenarian turtle are virtually indistinguishable from those of its teenage counterpart."[1] He goes on to say, "Turtles don't really die of old age" and "if turtles didn't get eaten, crushed by an automobile or fall prey to a disease ... they might just live indefinitely."[2] Likewise, there are trees on Earth that have been around thousands of years, like some of the Great Basin Bristlecone Pines that first sunk their roots into the Nevada soil almost 5,000 years ago.

As astonishing as it may sound, it is also true that when life began on Earth, in its exclusively cellular form, it *was* immortal and remained so for more than a billion years. "Many single-celled organisms *may* die, as the result of accidental starvation; in fact the vast majority do," writes cellular biologist, William R. Clark. "But there is nothing programmed into them that says they *must* die. Death did not appear simultaneously with life. This is one of the most important and profound statements in all of biology. At the very least, it deserves repetition: *Death is not inextricably intertwined with the definition of life.*"[3] In his 1996 book on the origins of death, Clark reminds us that cells reproduce by dividing, that neither is—and both are—the original, and they don't leave a corpse behind in the process. Death came later, with the advent of more complex, multicellular creatures, at which time some cells became senescent, meaning their ability to replicate began deteriorating over time.

Today humans, like most but not all multicellular creatures on Earth, are considered biologically mortal because our cells are prone to senescence. As we age, that is, they stop working like they're supposed to, leading to numerous age-related health issues, at least one of which will eventually kill us. Although this is so, that enough senescent cells eventually lead to fatal illnesses, we should keep in mind a growing number of scientists don't believe cells naturally become senescent. *Old age*, as we call this condition, is itself a disease, not an inevitability. Cells stop functioning and reproducing properly because of illnesses, and illnesses can be treated and often cured. In *Lifespan: Why We Age and Why We Don't Have To*, David A. Sinclair, professor of genetics at Harvard Medical School, says, "There is no biological law that says we must age. Those who say there is don't know what they're talking about."[4]

There are, of course, other reasons we die: death by predation, by injury, by inhospitable environments, and by other kinds of diseases. But even if none of these things kill us, humans eventually die because too many senescent cells lead to what biohacker David Asprey calls the *Four Killers*: heart disease, diabetes, Alzheimer's, and cancer. "Suffice it say that if you don't die in a car crash or from an opioid addiction, chances are that one of these Four Killers is going to drain your life and your energy (and your retirement fund) before you die in a hospital."[5] But Asprey isn't willing to settle for this inevitability, and has set a goal of living until he's at least 180. Sinclair, who says there is no upper limit, is equally convinced those of us alive today will be around a lot longer than most of us ever imagined. "Prolonged vitality—meaning not just more years of life but more active, healthy, and happy ones—is coming. It is coming sooner than most people expect."[6]

If it weren't for the diseases that kill us as we grow older, caused by diseased cells that stop functioning like they're supposed to, it's estimated the average human lifespan would be about 6,500 years. As Byron Reese writes in his book, *The Fourth Age*, "That's how long it would take for some freak accident to befall you, such as a grand piano falling out of a window and landing on you. In such a world, death would be even more of a tragedy, since an accidental death wouldn't just shave forty years off of your life, but four thousand."[7]

Today the average human life is a measly 79 years compared to that, but it's up from 47 years at the start of the 20th century. This increase is due largely to improvements in sanitation and medicine. "But it's a fact that nothing has changed biologically in the last century," biotech expert, Jim Mellon says. "If you take someone out of 1900 and put them in today's environment, they'll live just as long as we do."[8] In his 2016 book, *Juvenescence*, the opposite of senescence, Mellon argues that human longevity is in the process of taking a major leap forward. "In a nutshell," he says, "we believe that it is possible to extend average human life expectancy significantly just by using today's technology, to within a decade or so of today's current 'hard' ceiling of about 115 years."[9] If he's right, that means adding another 35 to 45 years to the average human lifespan within the next ten years. "Nothing's changed to our fundamental biology," Mellon continues, "but today we're on the cusp of a major change. The biological engineering of humans, the rearrangement of our atoms and molecules to effect longer lives is with us. There are human trials going on at the moment. This is not science fiction."[10]

In 2011, for example, a 29-year-old member of my congregation in Louisville, Kentucky, had been diagnosed with stage-4 melanoma that had spread to his lungs. He was told he had only 8 months to live. His family, needless to say, was devastated. Because he was young and otherwise healthy, and had nothing to lose, he was approved to undergo a dangerous experimental treatment in an attempt to stimulate his own immune system to recognize and destroy his cancer. He's still alive, has since married, has a child, and is enjoying life. The same year he was told he was both incurable and cured, the FDA approved a vaccine for the treatment of metastatic melanoma, based partly on his success. Since then, in just a few short years, the Agency has approved immunotherapy treatments for nearly 20 kinds of cancer. What's remarkable about these cancer treatments, which are still in their infancy, is that they fight the disease by altering a patient's own biology.

Between 2012, when Jim Mellon wrote *Cracking the Code*, about the biotech revolution, and 2017, when he wrote *Juvenescence*, he likes to point out that we've not only developed these immunotherapy treatments for cancer, but also Artificial Intelligence (AI) and gene editing, both of which have enormous medical application, and we've even cured Hepatitis C. *What will happen in the next six years?*

Mellon predicts gene editing will eventually inculcate geroprotective genes into the general population. These are genes that already allow some humans to naturally live much longer than most. In the meantime, longevity researchers are also focusing on small molecules, stem cells, and organ regeneration. Attempting to preempt age-related deterioration and ailments with compounds, supplements, and diet in this way is what Asprey means by *biohacking*, which, it turns out, even my spellcheck knows about.

Dr. Joan Mannick, the founder of the biopharmaceutical company resTORbio, points out that aging is a biological program that can differ radically in species that are otherwise very similar. The steamer clam, for example, lives about 28 years compared to the ocean quahog, a similar clam, that lives up to 220 years. A painted turtle has an 11 year lifespan compared to a Galapagos land turtle's span of 193 years. And a common mouse lives about 3 years compared to its cousin, the naked mole rat, that lives 28 years. Lifespan, it would seem, can be rather arbitrary. The difference, Mannick points out, has to do with TORC1 production, a protein complex that's active while eating and inactive while fasting when we're young. As we grow older, however, TORC1 remains active all the time and is associated with the development of age-related health problems. It's not as present in those species, like the ocean quahog, Galapagos land turtle, and naked mole rat, that tend to live longer. Inhibiting it, according to resTORbio, "has been observed to prolong lifespan, enhance immune function, ameliorate heart failure, enhance memory and mobility, decrease [body fat] and delay the onset of aging-related diseases in multiple animal studies." Mannick says her company is initially working to impact respiratory infections, a leading cause of death in seniors. During Phase 2 trials involving 900 patients, they've already achieved a 40 percent reduction in respiratory infections.

Samumed is another company specializing in regenerative medicine by creating treatments that make use of the WNT pathway, which signals the regenerative properties of cells in all creatures. WNT is abundant when we're young, but not so much as we age, which is why bones heal easier, our joints don't wear out, and we don't go bald when we're kids. Samumed has already had success growing new cartilage in bad knees with a single injection of the WNT pathway. Once this treatment is perfected and approved by the FDA, it could mean an end to the need for invasive joint replacement surgeries.

When overactive, the WNT pathway can also cause cancer by overproducing cells. So Samumed has developed a treatment to also inhibit WNT, which is currently in Phase 1 trials. The company was allowed to treat a 30-year-old woman with it, under the Compassionate Use Act, because she had terminal pancreatic cancer, was down to 70 pounds, and was sent home to die. After a year of treatment, her cancer was gone, she weighed 130 pounds, resumed athletics, and was living a normal life again. Samumed is also working on a treatment for Alzheimer's. It was given to an 80-year-old woman, also under Compassionate Use, who had become bedridden, nonresponsive, and unable to walk, eat, talk, or recognize her family. Within three months of treatment, she could do all those things again. Samumed has six treatments for other age-related conditions currently in Phase 1 human trials, as well as treatments for osteoarthritis, and male pattern baldness in Phase 3 trials.

Celularity is a medical company specializing in stem cell research and regenerative treatments. Stem cells that replenish whatever cells our bodies need when we're young, exponentially decline as we age, sometimes leading to cardiovascular, pulmonary, cognitive, or cancer diseases that 80 percent of people over age 65 have at least one of. In their research, animals that have had their stem cells replaced live 40 percent longer than their counterparts. Celularity is using placental stem cells, which are unlikely to be rejected by their hosts, to develop treatments that will actually reverse the aging process and help keep us active and healthy no matter how old we are.

Yet another company, Elevian, is developing treatments based upon studies showing animals "transfused with the blood of young animals, experience regeneration across many tissues and organs. The opposite is also true: young animals, when transfused with the blood of older animals, experience accelerated aging."[11] This is so, the company says, because of a molecule known as GDF11, or "growth differentiation factor 11," which becomes scarce as our cells undergo senescence. GDF11 supplementation "accelerates skeletal muscle repair, improves exercise capacity, improves brain function and cerebral blood flow, and improves metabolism."[12] In 2014, *Science* magazine named Elevian's findings one of the top 10 scientific breakthroughs of the year. Two years later, a similar drug produced by another company, which actually reduces senescent cells, was named *the* breakthrough of the year.

Some might think prolonging life isn't such a good thing, despite the survival instinct exhibited in most living creatures. They would argue it's only natural for us to eventually die and that it's Frankensteinish to interfere with Mother Nature. Others consider our aversion to death to be against their religious or spiritual beliefs. Whether valid or not, such arguments have little bearing on the issue given that we are already in the process of prolonging life and have been for more than a century. It will continue so long as there are those who don't wish to become infirmed and eventually die.

What can be more spiritual than the possibility of living longer and healthier lives, and what's more natural than the desire of any creature to survive? There's a scene in the 1985 film, *Cocoon*, about a group of people in a retirement community who discover a swimming pool with rejuvenating powers. Toward the end of the film, one of the aliens responsible for the technology invites some of them to leave Earth with him and go where they won't have to worry about dying or aging ever again. "You would be students of course," the alien says, "but you'd also be teachers. And the new civilizations you would be traveling to would be unlike anything you've ever seen before. But I promise you, you will all lead productive lives."

"Forever?" asks Ben Luckett, memorably played by Wilford Brimley.

"We don't know what forever is," the alien replies.

Ben then turns to Mary, his spouse, who seems reluctant. "So you think it's like Bernie said? We're cheating nature?"

"Yes," Mary says.

"Well, I'll tell ya, with the way nature's been cheating us, I don't mind cheating her a little."

I suspect those among us who are young at heart—young souls trapped in aging bodies—can relate, and, given the opportunity to feel better and be as active as they'd like, would also say yes to what may currently sound like alien technology. It's a question we must all begin thinking about because these advances in science and medicine are happening, accompanied by many ethical and social ramifications.

Currently, the fastest growing demographic everywhere on Earth is 65-year-olds, 80 percent of whom, again, have at least one age-related illness. And with the global population decline causing most countries to now be at or below their necessary replacement birthrates, there are going to be fewer young people to help replenish the finances and workforce needed to care for a society dominated by elders. One solution is to dismiss anti-aging technologies as unnatural and selfish, condemning everyone to die around a "normal" age, which was around 30 years until the 18th century, 47 in 1900, and almost 80 today. How would we feel today if a hundred years ago our society had said, "Sorry—

sanitation, antibiotics, and vaccinations aren't natural. You'll just have to die before you're 50."

It seems a better, more humane, and more likely solution is to reverse the impacts of aging, eliminating rising eldercare costs in the process. According to a *60 Minutes* story, Medicare annually pays more to U.S. doctors and hospitals in the last two months of patients' lives than the entire "budget for the Department of Homeland Security, or the Department of Education."[13] More than $125 billion Medicare dollars, about a quarter of its annual budget, goes to 5 percent of its recipients during their last year of life. And that report was in 2010. According to the Kaiser Family Foundation's website, the number of Medicare recipients has increased from 46 million to more than 60 million since then. Imagine what we could do with this amount of money if it weren't being spent on age-related illnesses, and wonder what we'll do if such a program runs out of money because it's overwhelmed by the number of people needing it, yet no longer has enough young people working to help replenish it. If we can cure their illnesses or prevent them from ever occurring to begin with, we should, because, as Wilford Brimley also famously said, "It's the right thing to do."

Living longer and healthier also requires us to think more deeply about our moral obligations to younger people.

How do we continue to live long, healthy, productive lives without robbing them of the opportunities to do the same? If older folks aren't retiring, where will some of them find jobs? And if we're still healthy enough to work but don't, what will we do with our time? What will society look like when people are active for more than a century, and probably much longer as technologies continue to exponentially advance? More importantly, what do we want it to look like? If we're planning to reengineer human biology, we ought to simultaneously plan to reengineer how human society is going to work. Let's not just wait and see.

The times we're living in are as exciting as they are uncertain and give us as many reasons for dread as for hope. That *the future remains uncertain* is a truth that will likely never change. But humanity is also on the cusp of a transformation once the stuff of pure science fiction. We can still hardly imagine what wonder will burst free from the cocoon of disease and death that has so long encased us.

---

[1] "All but Ageless, Turtles Face their Biggest Threat: Humans," by Natalie Angier, *New York Times* (Science), December 12, 2006.
[2] Ibid.
[3] Clark, William R., *Sex and the Origins of Death*, Oxford University Press, New York, NY, 1996, p. 54.
[4] Sinclair, David A., *Lifespan: Why We Age and Why We Don't Have To*, Atria Books, New York, NY, 2019, p. xxiii.

[5] Asprey, David, *Super Human*, Harper Wave, New York, NY, 2019, p. 4f.
[6] Sinclair, ibid., p. xxii.
[7] Reese, Byron. *The Fourth Age: Smart Robots, Conscious Computers, and the Future of Humanity*, Atria Books, New York, NY, Kindle Edition, 2018.
[8] Abundance 360 Conference, Module 6: Longevity and Vitality, January 2019.
[9] Mellon, Jim, *Juvenescence: Investing in the age of longevity*. Harriman House. Kindle Edition, 2017.
[10] Abundance 360 Conference, ibid.
[11] www.elevian.com
[12] Ibid.
[13] http://www.cbsnews.com/news/the-cost-of-dying-end-of-life-care/

SEVENTEEN

THE HOUSE OF HUMANITY

Economics isn't merely about how we manage money, but about how we manage all our resources, including education, employment, healthcare, housing, the environment, political power, our time, and much more. The thought that economics is limited to financial matters is a modern fallacy that leads too many of us to believe most of our problems are rooted in not having enough money and can be solved by having more money.

Economics was considered a religious matter long before becoming the exclusive turf of financial experts. *Economy* is comprised of two ancient Greek words common in the Christian scriptures, *oikos*, meaning, "a home," or, more specifically, "an inhabited house," and *nomos*, meaning, "custom," or, "law." Together, *oikonomos* is literally translated, "manager of household affairs," or more simply, "steward." As a verb, *oikonomia* refers to the stewardship, oversight, and management of household affairs.

The meaning of "house," furthermore, *oikos*, doesn't refer to a physical structure, but to those "dwelling in," or "inhabiting" a place. A house is a family not a building, which is why another version of the word, *oikeios*, is translated to mean, "kindred," or "belonging to a household." *House* comes from the Old Germanic word, *hus*, meaning, "family." So, economics isn't about financial institutions, systems, and structures, but about those living within these institutions, systems, and structures. In his book, *The Sane Society*, Erich Fromm laments the loss of this "most important of all" attitudes, which he says, "had determined the life of [humanity] for centuries ... the principle that society and the economy exists for [humankind], and not [humankind] for them."[1]

In the past, families also referred to themselves as a *House*, like the House of David, or the House of Usher. The Hebrew scriptures refer to "both houses of Israel,"[2] indicating its northern and southern kingdoms: two entire societies considered part of one household. Today, in a globalized society inhabiting one planet, living under one atmospheric roof that shelters us all, the success of the economy should be determined by the welfare of all people, which necessarily includes the proper stewardship of the Earth, its resources, and all its creatures. An economy that works well for only a few people, or concentrates only on making one country great, or creates jobs and industries by harming and destroying the environment, is a failed economy. Today, a successful economy must adequately manage the needs of everyone and everything under the Sun.

So the good stewards among us are looking ahead, trying to discern what the future has in store for the House of Earth and its Human Family and to see what must be done to assure all is managed well so that all are well. Ever since Abraham Maslow introduced us to his hierarchy of needs, we've known that for a society to work, for its economy to be considered successful, it must, at the very least succeed in meeting the basic needs of everyone under its roof, including food, water, air, rest, safety, and security. Once these are met, individuals are free to concentrate on fulfilling their needs for belonging, self-worth, and meaning in their own way.

This doesn't mean our governments must provide all these things to us for "free." It only means that whatever economic systems are in place must work to achieve these ends, or else they should be considered failed economies. A capitalist society that generates a lot of financial resources that end up in the hands of too few, at the expense of the environment, and that results in too many people living in poverty and struggling just to meet their most basic needs, is a failed economy. A communist or socialist economy that provides the very basic needs for everyone, but does so by restricting the freedoms of individuals, suppressing their need to fully unfold as creative, productive, unique, happy, self-actualized persons is a failed economy. How then are we to have a successful economy, signified by the wellbeing of everyone and everything, that also values individual autonomy?

In the runup to the 2016 U.S. elections, Bernie Sanders achieved unprecedented popularity by suggesting things like universal healthcare and publicly funded higher education. In 2019, other would-be Presidential candidates included these in their platforms, in addition to other new entitlements, like free childcare. One candidate, Andrew Yang, even proposed a guaranteed income of $1,000 a month to every U.S. citizen over the age of 18. Yang's website points out the idea of a universal basic income (UBI), which he calls a Freedom Dividend, has been around since the Renaissance and was promoted by people like Thomas Paine, Martin Luther King, Jr., Richard Nixon, Milton Freidman, and, most recently, Barack Obama, Elon Musk, and Mark Zuckerberg.

Musk and Zuckerberg, in particular, recognize emerging technologies, especially robotics and artificial intelligence, are poised to cause a massive disruption in the workforce. In the near future we simply aren't going to need or have enough jobs to keep everyone employed. Many forward thinkers, including Andrew Yang, want to address this coming certainty before it becomes a global humanitarian crisis. So there's a lot of talk these days about establishing a UBI, along with serious conversations about what a successful society might look like if there isn't much need for people to work, at least not in the traditional sense.

One of its strongest proponents, the late economist and futurist Robert Theobald, once lived in Spokane, Washington, where he was a friend to many in the congregation I've served, and often spoke about UBI from its pulpit. In his 1966 book, *The Guaranteed Income*, he predicted the then-budding computer technologies would eventually cause people in the middle-income level to drop to the poverty level.

As the title of his book indicates, Theobald proposed that securing a robust middle-class requires establishing a guaranteed income for everyone, especially in light of advancing computer technologies. He believed this would empower people to, in his words, "break out of the consumption race," by which he meant the need to constantly earn enough money just to pay for their immediate needs and take care of their monthly round of bills. A UBI would, he thought, make us human again and not just consumers. He also believed a guaranteed income would grant individuals the time and resources necessary to focus on what they feel is truly important, instead of using most their energy just to make ends meet. "We can then set free enormous potential energy and imagination to solve our urgent problems,"[3] he said. More people would be willing to take risks by being creative and innovative. Finally, Theobald believed that not having to worry as much about attracting job-producing industries would free entire communities to decide what kind of city or town they want to focus on being: an education city, a garden city, a recreational city, an art city, and so on.

    This all sounds pretty good, but also raises some practical questions, like how much will it cost and how are we going to pay for it? Andrew Yang addressed these questions and others during his campaign, but some of his responses weren't wholly convincing. He can't guarantee a UBI won't cause inflation, even though he argues it won't. Making an extra grand a month could easily and quickly become less valuable that it sounds. Nor did he adequately explain how $12,000 a year would mean the same to someone living in Seattle or San Francisco as it does to a person living in a smaller city or town.

My bigger concern with the notion of a UBI, however, is that it boils all our economic problems down to a lack of enough money and their solutions to having more of it. I would further argue something like this has already been tried and failed. When President Franklin D. Roosevelt signed the Federal Labor Standards Act into law in 1938, guaranteeing every worker would earn at least 25¢ an hour, he did so, he said, to maintain "a minimum standard of living necessary for health, efficiency and general wellbeing."[4] Today, more than 80 years later, the U.S. Federal minimum wage is $7.25 per hour, a 2800% increase. During this same period inflation rose only 1,717.61%, which would make the equivalent of 25¢ today only, $4.52.[5] That's about 40% less than $7.25. Over this same period, however, the minimum wage rate increased 1% faster than inflation, making $7.25 worth 60% more now than 25¢ was in 1938.

But this probably says more about how little 25¢ meant back then than it does about how much $7.25 means now, at least in the U.S. Today, even if it were raised to $15 an hour, a *full-time* minimum wage job would pay just

$31,200 annually, before taxes and benefits. In 2018 the average rental cost for a one to two-bedroom apartment in the U.S. reached $1,405 per month.[6] That leaves $14,340. Deduct the average cost of medical insurance, $440 a month for individuals and $1,168 per family, plus an average $8,232 annual deductible,[7] taxes and other benefits, and even a single person has nothing left to buy groceries or clothes with, let alone to make a car payment, or pay for auto insurance, utilities, life insurance, investments, retirement funds, savings, emergencies, or for vacations and just having fun. Even so, many in our society are advocating for a $15 per hour minimum wage, which simply isn't adequate to provide our most basic physiological needs.

If a $15 an hour minimum wage is achieved, then all we'll really guarantee is that everyone in the U.S. lucky enough to find a full-time job can at the very least struggle every day just to survive. What happens when there's something equivalent to another stock market crash, or something like the subprime lending scandal that came to a head in 2008? After 83 percent of the risky loans at the time, about 12 million of them,[8] were approved by unregulated private lenders, stricter banking regulations were instituted across the board, making it now more difficult for many borrowers to qualify for home loans from reputable lenders.

This has led to greater demand for rental properties, causing their prices to skyrocket, making it more expensive to rent a home in the U.S. than paying a monthly mortgage. Yet those who can no longer qualify for a home loan have no choice but to pay these exorbitant rental prices or else live in squalor, shelters, or on the streets.

The overall decreases in U.S. homeless rates in recent years are being offset by their increases in some of the nation's most populated and, therefore, most expensive places to live: the largest increases being among veterans, the chronically homeless, unaccompanied minors, and young adults.[9] Many workers, including police officers and teachers, can no longer afford to live in the same communities they serve and must drive several hours a day getting to and from work. Some long-established businesses in major cities are even closing because they can't find enough low-income wage earners living close enough to work for them.

The point is, the meaning of money, including $1,000 a month, is volatile and only one economic disaster away from becoming meaningless. We need to cease conflating economics with money and recall that economics is about human wellbeing, which, again, necessarily includes proper care for our environment and other creatures.

So instead of talking about UBI, let's start talking about providing a basic quality of life (BQL). This, after all, is what a universal basic income is supposed to achieve, to help everyone fulfill their basic needs. Instead of talking about giving everyone enough money to buy their basic needs, let's talk about directly providing their basic needs. Let's imagine a society that considers having a home, healthy food to eat, a quality education, and good healthcare, to be basic signs of a successful economy, and not consumer products only those with high paying jobs can secure.

Sounds good, but how on Earth can any country pay for all that? As far as I'm concerned, the answer isn't giving people more money. It's about lowering the tremendous costs of housing, higher education, and healthcare. Even if today's minimum wage is 60% more valuable than 25¢ was in 1938, it's still not enough when considering the overwhelming expense of these three essential needs. Yet were it not for these exceptions, think about all the things most of us do have access to these days that even the wealthiest couldn't have afforded 80 years ago. Back then running water and indoor toilets were still luxuries. 50 years ago, even millionaires didn't own personnel computers. Today many of the poorest among us have access to computers and own smartphones, the most powerful and sophisticated technologies in human history. So, in some ways, $7.25 is worth a whole lot more today than we tend to think.

This is so because of Moore's law. To oversimply, it means as computer technologies, in particular, evolve, they exponentially become smaller, more powerful, cheaper, and widely accessible. Today there are free apps on the smartphones in our pockets and purses that provide us with technologies that would have cost hundreds of thousands of dollars and filled rooms in our homes just a few decades ago. This means Moore's law also leads to the democratization and demonetization of technologies as they become abundant and free, or nearly free. So far, this trend has proven true with the exception of these three vital areas—housing, higher education, and healthcare. Despite all our advances in technology, the rising prices of these crucial commodities make them less accessible for most people, not more accessible. But this too, some predict, is about to finally change thanks to current advances in technology and the application of Moore's law.

Right now, the technology exists and is being used in places around the world to 3-D print houses in about 24 hours for around $4,000 to $10,000 each. That's right now, even without the technology being in widespread use and while it's still in its early stages of development. At this price, printing houses is going to become cheap very quickly, enough so that any society can easily afford to make home ownership a right for its citizens.

Likewise, education is undergoing massive disruption thanks to AI tech that's already tailoring it to the specific needs of students, along with companies that can train their employees to meet the specific demands of their jobs with far greater precision than any University can. *Udacity*, for example, now educates and trains employees in areas like AI, computer programming, data science, autonomous systems, robotics, and cloud computing, for companies like AT&T, Google, Amazon, and IBM, among others. It says its average enrollment is about 5 months, the average tuition is about $1,300, it has students in 190 countries, 84% of graduates get a job within six months, with an average $24,000 annual pay increase, and if they don't get a job in six months, they get their full tuition returned. Given some of the frustrated online reviews I've read, it does seem companies like Udacity could still benefit from having some more traditional human interactions, but as they continue to improve their services, and continued advances in AI, the age of indenturing college graduates to enormous student loan debt just for the right to work could abruptly end.

Finally, new medical technologies are also disrupting the healthcare industry. The wealth of medical knowledge will soon become accessible at our fingertips with apps on our smartphones. There are already companies that, for a small fee, provide us access to real doctors and prescriptions right on our devices, a service many health insurance companies now make available. This is great news considering half the people in the world never see a doctor during their entire lives, yet, in most countries, between 70 and 100 percent of us have smartphones.[10] In the U.S., where healthcare costs remain high, 80 percent of Americans delay seeking preventive care,[11] and the average American only sees a doctor four times a year.[12]

Stanford University has recently developed a chip that inserts into a smartphone that, with only two drops of blood, tests for sensitivity to 96 allergens at once. There is now technology from *AliveCor* that, for about $150, turns a smartphone into a six-lead electrocardiogram (ECG) machine that can be used at home in the U.S., or in some of the remotest communities in the world. *Healthy.io* has developed smartphone urinalysis kits that instantly and conveniently check for issues like UTI, kidney disease, proteinuria, and other conditions for less than $15, the results of which can be instantly sent to one's doctor. Between technologies like these and AI that can make more reliable diagnoses based on millions of data points, digital medicine is upon us and is about to transform our expensive sick-care systems into genuinely affordable healthcare.

These three crucial and costly areas—housing, education, and healthcare—have recently entered the realm of exponentially advancing technologies, which will soon make them affordable and accessible to everyone. This is going to be especially important in a world where human workers are displaced by robots and artificial intelligence. More importantly, our social architects and community leaders, whether they are conservatives or progressives, should be aware of and working toward these changes because they are going to help us achieve our social goals regardless of our politics, or whether our emphasis is funding social programs or reducing deficit spending.

For our economies and societies to be successful, they must result in a BQL for every person, which doesn't mean simply throwing more money at our problems. In fact, I predict, through the democratization and demonetization of things, money itself will eventually become meaningless in our societies. But whether we use money or not, being good stewards of our home, the Earth, and caring for the House of Humanity means finally making sure everyone under the Sun has what they need to live a meaningful and productive life. Providing food, shelter, clothing, clean air and water, along with safety and security, and access to the kind of learning and good health that help make our lives worth living, as we seek meaning and strive to fulfill our purposes, can be the only measure of a society's economic success.

---

[1] Fromm, Erich, *The Sane Society,* Henry Holt & Company, New York, NY, 1955, p. 85.
[2] Isaiah 8:14
[3] https://www.youtube.com/watch?v=BUTwZUHJRTI
[4] https://www.minimum-wage.org/articles/history
[5] This calculation is based on the U.S. Bureau of Labor Statistics' figure of a 3.6% average rate of inflation since 1938. [25*(1+3.6%)^82=4.52]
[6] https://www.rentcafe.com/blog/rental-market/apartment-rent-report/rentcafe-apartment-market-report-june-2018/
[7] https://www.chealthinsurance.com/resources/individual-and-family/much-health-insurance-families-cost
[8] Denning, Steve, "Lest We Forget: Why We Had a Financial Crisis," *Forbes* Senior Contributor, Forbes.com, November 22, 2011.
[9] https://endhomelessness.org/homelessness-in-america/homelessness-statistics/state-of-homelessness-report/
[10] Harrison, Guy P., *Think Before Your Like*, Prometheus Book, New York, NY, 2017, p. 17.
[11] "Why Americans are putting off doctor visits: Zocdoc," CNBC, Published Tue, Jun 23, 2015, 2:40 PM EDT.
[12] https://www.commonwealthfund.org/press-release/2015/us-spends-more-health-care-other-high-income-nations-has-lower-life-expectancy

EIGHTEEN

LEAVING EARTH

Life leaves or dies trying. This is one of the principles essential to the process of evolution. All life is invasive. It's never satisfied with the way things are. It's always on the move, trying to become something more. When our Universe began 13.7 billion years ago it was smaller than the tiniest of seeds, yet somehow burst beyond its confines and has been expanding ever since. It's still pushing its own limits and boundaries. Ours is a transcendent universe.

This is true, not only of what has become its incomprehensible expanse, but of everything about it and within it. When it began, the Universe had only one element, hydrogen. Eventually this lone element formed immense communities of hydrogen clouds, becoming so heavy they collapsed. Their internal temperatures became hot enough from the pressure to form the first generation of stars, fusing hydrogen particles to create a second element, helium. When the pressure within these stars became too intense, they, like the Universe itself, began expanding, either by casting off their outer layers or becoming supernovae, making the entire Universe something more than it had been before.

This same process has been continuing for eons, our transcendent Universe forever expanding and becoming increasingly complex, becoming more than it's ever been before. It wasn't long after its own birth that the Universe became complex enough to form the first planets. But these were gaseous planets composed mostly of hydrogen and helium balled together by gravity. We know it wasn't until harder elements, like iron, magnesium, and silicon were formed, that it became complex enough to form rocky planets, including our Goldilocks planet 4.5 billion years ago. A billion years later the Earth's transcendent chemistry became complex enough to form life. These most ancient of our ancestors, the prokaryotes, were single-celled microorganisms so simple they didn't even contain a nucleus. But they weren't content to live as such simpletons. After another billion years had passed, they evolved into eukaryotic cells, complex enough to contain small organelles, including a nucleus and mitochondria.

They still weren't satisfied. Our eukaryotic ancestors could not contain themselves. Another billion years after our *simple* unicellular ancestors transcended into *complex* unicellular beings, they converged to form the first multicellular organisms—boneless, brainless globs of eukaryotic algae. Living in such close proximity, naturally communicating genetic materials, proving fatal in some cases and beneficial in others, which, through this process of natural selection, evolved into sexual reproduction. After about 600 million more years, this first sexual revolution led to the emergence of simple animals like sponges, fungi, and coral. Though sturdier and more complex, and capable of sexual reproduction, locomotion, and digestion, these first true animals on Earth still weren't content to remain where they were or what they were.

Their continued discontentment with staying put caused the Cambrian Explosion, and life began diversifying faster than ever. Some organisms even dared venture from the life-sustaining oceans onto the hot, barren land. Many died in the process of getting trapped in the unbearably hot shallows and corrosive atmosphere. But some, presumably with just the right mutations, learned to survive the heat, and later to directly consume sunlight, rather than digesting other organisms. They became the first plants. Some plants later found their way into the seas, while some ocean creatures followed the original plants onto shore.

From this perspective, what Darwin named *evolution* can also be considered an ongoing process of transcendence, of everything in the Universe, including its elemental, chemical, and biological structures, continuing to move beyond where and what they are. We live in a transcendent Universe and everything in it moves toward transcendent states.

Part of *transcendence*, a word that literally means, "climbing over," or, "surmounting," includes transcending location. Life likes to get around or dies trying. All species are invasive species. Plants move about by letting the winds or other creatures carry their pollen and seeds to new places, sometimes adapting along the way. Other animals move about on their own, by land, air, or sea, also adapting themselves to new environments via natural selection. Among mammals, the human animal has been most successful at this. We've migrated and adapted to living almost everywhere, from the hot and humid savannahs to the frozen arctic tundra, from lush green forests and jungles to barren deserts, and have dipped ourselves deep into oceans and blasted ourselves far into outer space. We've accomplished all of this, transcending our natural environments, despite the limitations our bodies have enduring extreme heat, frigid cold, watery environments, and what should be certain death anywhere outside the Earth's atmosphere.

But we haven't only been great at transcending our natural environments. Human beings have another exceptional power, unlike that of any other animal we know of, the ability to liberate information from the unforgiving confines of our genes. Certainly, other animals are capable of learning new behaviors, some of which others of their kind might mimic. But most nonhuman animals, as far as we know, are born with most of the knowledge they need to survive. They have genetic knowhow expressed through instinct. Human beings, on the other hand, must learn the behaviors necessary to navigate and survive our ever-changing environments and cultures by being educated for many years after we're born. As our civilizations advance, each generation must learn new skills and knowledge.

In this way, humans represent a major divergence in evolutionary history. Through our species, information has begun to transcend the limitations of biology. Through us, information can now be exchanged inorganically, simply by exchanging nonphysical constructs known as *ideas*. Whether it is by listening to a sermon, reading the paper or a good book, or talking with a friend on the phone or over a cup of coffee, a near miracle on Earth occurs. Inorganic nonphysical information is exchanged between two organisms through memes, not genes. It's what makes our species, despite its many downfalls, beautiful and special. We represent a transition of life from purely physical to nonphysical states. Maybe this is why so many of our religions imagine life beyond the physical, because these mythologies reflect some natural longing of life itself to transcend its physical constraints, its need to invade new dimensions of existence.

Pierre Teilhard de Chardin, the beloved Jesuit priest, mystic, and scientist who fully embraced evolution, a rare combination, envisioned it eventually resulting in information taking on a life of its own, until our planet itself becomes, in his terms, the "thinking Earth."[1] In 1947 he wrote, "we must enlarge our approach to encompass the formation, taking place right before our eyes ... of a particular biological entity such as has never before existed on earth—the growth, outside and above the biosphere, of an added planetary layer, an envelope of thinking substance, to which, for the sake of convenience and symmetry, I have given the name of the Noosphere."[2]

A generation later, in 1982, Peter Russell, a theoretical physicist and computer scientist by training, published his bestselling book, *The Global Brain: Speculations on the Evolutionary Leap to Planetary Consciousness*. In it he promises, "We shall see that something miraculous may be taking place on this planet, on this blue pearl of ours. Humanity could be on the threshold of an evolutionary leap, a leap that could occur in a flash of evolutionary time, a leap such as occurs only once in a billion years."[3] He was, of course, talking about what in 1982 was soon to become known as the Internet, "electronically based telecommunications networks (telephones, radio, computer links) are like the billions of tiny fibers linking the nerve cells in the brain."[4] Whatever we call it, the Noosphere, the Global Brain, or the Internet, this notion that the Earth is developing a mind of its own isn't new, and today isn't even farfetched.

What I like about de Chardin's description of it, as something "outside and above the biosphere, a new planetary layer," is the suggestion the boundaries of the Earth are now expanding outward, deeper into space, with this biologically liberated layer of invasive information. The Internet may not exactly dwell in outer space, though it is a world wide web of increasing information systems surrounding the globe, analogous to the formation of a neural network. Just as the

Earth has an atmosphere, and a stratosphere, it now has a noosphere, a sphere of digital knowledge. This network is now rapidly growing beyond the mere amassing of humanity's collected knowledge by also connecting all our technologies. It is turning into an Internet of things. Automobiles, cellphones, streetlamps, cameras, even entire buildings, and everything in them are now becoming linked together in the Cloud.

Some are concerned about what this global brain, this thinking Earth, might mean for us mere humans, and there are legitimate reasons to be cautious. But we should also remember this new way of being is an extension of our transcendent selves. Like Zeus giving birth to Athena from his head, this budding intelligence is our brainchild, born not of our genes but of our memes. The connected global mind is a new power, a new ability, that we are evolving for ourselves. The crucial part is not becoming mindless cells in its vast network, controlled by the extreme groupthink that the pile-on culture is attempting to force upon us through today's social media. We must remain differentiated cells in the network of thinkers and things, not undifferentiated mindless automatons.

Meanwhile, as liberated information continues evolving a mind of its own, our invasive species continues transcending the boundaries of our natural environment by venturing deeper into space. On Oct. 4, 1957, humanity launched Sputnik, its first ever satellite, into space. A month later, Sputnik II carried its first living passenger into space, a dog named Laika. In 1958, Explorer I became the first U.S. satellite successfully launched into space. All of these remained in Earth's orbit, but in 1959 the Soviets sent Luna I to the moon, an unmanned spacecraft that missed its moonshot but still became the first human piece of technology to transcend Earth's orbit. That same year NASA deployed a couple of spy satellites, captured the first-ever photographs of Earth from space, and the Soviets launched a craft that made it to the surface of the moon, albeit a crash landing. In 1961 they launched a probe to Venus, which traveled for a week before contact with the craft was lost. That's the same year Yuri Gagarin became the first person in space, followed by Alan Shepard, the second human in space, only a month later. In 1962, the Soviets deployed another spy satellite and NASA succeeded in completing the first interplanetary flyby by sending a probe, the Mariner II, past Venus.

In 1964, the space race resulted in a successful spacecraft designed to carry two to three passengers into space, and in 1965 cosmonaut Alexei Leonov was the first person to spacewalk. That same year, Ed White, an American, became the second man to do so and the Mariner 4 succeeded in a flyby past Mars. In all, there were three different crewed missions completed by NASA's Gemini program between 1961 and 1966. In 1966, the Soviets successfully landed an unmanned craft on the moon. In 1968, the Apollo 7 and 8 missions allowed astronauts to orbit the Earth and the Moon, respectively, and in 1969, the Apollo 11 mission succeeded in placing the first man on the moon.

In 1972, Pioneer 10 became the first spacecraft to transcend our solar system. In 1973, Skylab, the first space station, was launched. In 1974, Mariner 10 flew over Mercury. In 1976, even as other nations were getting involved in space exploration, the Viking 1 landed on Mars, with Viking 2 already on its way. In 1977, Voyager 1 performed flybys past Jupiter and Saturn, even as Voyager 2 was heading for Uranus and Neptune. The same year the Soviets launched an orbital docking station. In 1978 the first GPS satellite was launched. In 1979, Pioneer 11 flew past Saturn. Since all these historic first-time events, we've launched the Mir space station, developed reusable space shuttles, mapped the surface of Venus, sent probes to Jupiter, launched the Hubble Space Telescope, explored asteroids, deployed privately owned space labs, launched direct TV satellites, and launched the

International Space Station where more than 220 astronauts from 17 different countries have visited and lived.

Today, the Voyager 1, which entered interstellar space in 2012, 12 billion kilometers from the Sun, represents the extent of our transcendence thus far. There are also nearly 5,000 satellites in orbit around our planet right now. In 2019, Elon Musk announced his company Starlink's plans to surround the planet with 12,000 satellites to provide instant, high speed Internet access to everyone everywhere on the planet. In addition to having an atmosphere, stratosphere, and noosphere around the planet, we also now have a growing *satesphere* (a sphere of satellites).

For many, much of this may seem foolhardy and unnatural, or unimportant compared to the many problems we must solve right here on Earth. But to me it seems as natural as the first cells on Earth emerging from their primordial soup, and simple algae adapting to live on dry land, and fish crawling out of the sea, and mammals crawling back in. Today, it's still difficult to imagine life transcending its biological confines, or for biology adapting to live outside Earth's atmosphere. But if we consider how far we've come already, recognizing since Sputnik was launched not so long ago we have already made enormous encroachments into outer space, including the addition of new planetary layers, a noosphere and satesphere. If it's true, as Peter Russell says, that we are amidst "a leap such as occurs only once in a billion years," it may be that our invasive species is beginning to climb out of its own primordial soup by adapting to and adopting the Universe itself as our new natural environment.

Ralph Waldo Emerson, the founder of Transcendentalism, once said, "We live in succession, in division, in parts, in particles. Meantime within [humanity] is the soul of the whole; the wise silence; the universal beauty, to which every part and particle is equally related; the eternal One."[5] As terrifying as leaving can be and venturing forward into the unknown is, as difficult as it is for us to change, the strange drive within us to surmount these fears is but the transcendent Universe itself calling us home, to be at one with all that is.

---

[1] de Chardin, Pierre Teilhard, *The Future of Man*, Harper & Row, New York, NY, 1959, 1964, p. 156.
[2] Ibid.
[3] Russell, Peter, *The Global Brain*, J.P. Tarcher, Inc., Distributed by Houghton Mifflin Company, Boston, MA, 1982, p. 7.
[4] Ibid., p. 32
[5] Emerson, Ralph Waldo, *The Writings of Ralph Waldo Emerson*, ed. Brooks Atkinson, Random House, Inc., The Modern Library, New York, NY, 1940, p. 262.

NINETEEN

THE TOOLMAKER'S PARADOX

Dualistic thinking is problematic because it causes us to draw lines where there aren't any. Just compare a map of the world to a photograph of it and you'll see what I mean. We behave as if the imaginary lines we draw in the sand are real and worth fighting over, even though they've barely scratched the surface of constantly shifting sands. But nobody is to blame for such dualism. Nobody invented it and convinced the rest of us it's how we should think. Differentiating ourselves from objects, others, and the environment is what enables us to become conscious. Those unable to make these distinctions remain unconscious and unaware. They are undifferentiated.

We perceive truth and reality through the lens of dualism, the lens of opposites and opposition—this is *this* and that's *that*. This is the meaning of Taoism's Yin and Yang mandala. These opposites are endlessly churning and turning into each other, neither with a discernable beginning or end, each always containing a spot of its opposite. It is through such duality we must, as conscious beings, interpret the world. Yet truth and reality are much larger and more whole than we are capable of perceiving or comprehending. If we don't recognize this, if we can't cope with cognitive dissonance by entertaining opposing ideas, we become rigid in our thinking, cruel in its execution, and go through life acting like halfwits. The mandala must be treated as a whole—the opposites are part of one unified reality.

Thinking and *feeling* is a common duality many accept without question. If one is "too smart," or "too logical," many presume one must be incapable of also having compassion or empathy for others. Others presume one who feels too deeply can't be thinking straight. But unless there's something terribly wrong, most of us are capable of both thinking and feeling. Deep thinkers are also deep feelers. Thoughts and feelings may, in fact, be one event. Thoughts incite emotional reactions, and we get in touch with our feelings by rationalizing them to ourselves or others.

Another duality is the notion that empiricists can't be spiritual. This reflects one of the most ancient, if not archetypal, dualities, the duality of Heaven and Earth. Either one is a gross, unenlightened, suffering, earthly creature, or one is a transcendent, high-minded, heavenly being. Yet scientists are empiricists. So are many philosophers. And you better hope your doctor is an empiricist, lest you prefer leaving your care to faith healers or the Fates! Anyone who is inspired by anything—and all of us are inspired by something—is spiritual, even those inspired by elegant science, nature, reason, math, and other "earthly" muses.

Disrespecting the spirit of another, which means not acknowledging it—because that's what *respect* means, *too see*—is also dehumanizing, which leads to the worst false dichotomy, the duality of *human* and *machine*. In his 1999 book, *The Age of the Spiritual Machines*, Ray Kurzweil says, "Twenty-first-century machines—based on the design of human thinking—will do as their human progenitors have done—going to real and virtual houses of worship, meditating, praying, and transcending—to connect with their spiritual dimension."[1] As astounding as this prediction is, that our machines will someday be spiritual, Kurzweil's provocative title alone is enough to disrupt the duality that we commonly take for granted exists between us and our machines.

Just try imagining if, instead of abstract Yin and Yang, we call the two sides of the Taoist mandala, *human* and *machine*, suggesting they are being born from each other, are turning into each other, and are so related that each always contains something of the other within itself. We could do the same with the other dualities I've mentioned and not be bothered. Say one side is *intellect* and one side is *emotion*, or one side is *matter* and the other *spirit*. This is a helpful exercise because it reminds us these opposites are within all of us and together make us whole. None of us is simply one or the other, either intelligent or emotional, either material or spiritual. We can both think and feel. We are both body and spirit. But the notion that humans are machines and machines can be human isn't so obvious and may even cause some to recoil in disgust.

My point here isn't to argue, as Kurzweil and others do, that humans and our machines are coevolving and converging: but that our dualistic thinking on the matter causes us to consider technology to be the very opposite of what it means to be human, and that those who overly use or appreciate technology must be a bit inhuman themselves. I began considering this dualistic paradigm recently when a good friend asked if I haven't strayed from the path of what I call my "North Star"—the humanistic ethic—since attending Singularity University.

My friend's question, which stems from the common misconception this *human* vs. *technology* duality leads to, is a good one, and deserves a thoughtful response. But it's not the first time I've been accused of disregarding the humanistic ethic in favor of machines. Several years ago, I had the opportunity to spend a day with a delegation from Afghanistan. Due to the U.S. war there, and to what they considered the unjust, unwarranted, and deadly preemptive war in Iraq, and because of the tremendous hurt and loss they were still experiencing because of my country's perceived heartlessness, they were initially rather hostile toward the first Americans some of them had ever met, including me. But only a few hours later, during our lunch together, we had already been around each other enough to recognize our common humanity. One of them told me, "We thought that you Americans had been around your machines for so long that you had become machines, without hearts. But now we know that you are human beings and have hearts just like we do." The belief in this duality is widespread. It's global. It's universal.

THE TOOLMAKER'S PARADOX

Many consider technology the opposite of humanity, and those who use it, a bit inhuman. When it comes to using technology to blow up our neighbors and their neighborhoods, I'd say there's some serious truth to this. What could be more inhuman than killing other humans? Yet anthropologists have also long distinguished toolmaking as our species' specialty, and *technology* is just another word for "toolmaking." Technology is, perhaps, the most human kind of undertaking. Fish swim, birds fly, and humans make tools.

So my answer to my friend's question is, *no*, I have not veered off course. Rather, I went to Singularity University in *pursuit* of my North Star. I went there because of my commitment to the humanistic ethic. Erich Fromm says the humanistic ethic "is based on the principle that 'good' is what is good for [humanity] and 'evil' what is detrimental to [humanity], *the sole criterion of ethical value being [human] welfare.*"[2] This means everything we do, every aspect of our societies, ought to be for the purpose of promoting human welfare and individual unfolding, including economics, politics, our treatment of the Earth and other creatures, and the technologies we develop. This, it so happens, is the explicit mission of Singularity University. It isn't merely about educating, inspiring, and empowering leaders to apply technologies, but to apply them for the specific purposes of addressing "humanity's grand challenges."

During its 2019 annual Summit, for example, SU's Executive Director Will Weisman began the event saying, "I'm here because like so many of you I believe in my core that an abundant world is possible in the not too distant future, and I want to do everything that I can to help bring that to fruition. To me that looks like a world where we feed everyone, where we educate everyone, where we shelter everyone, a world where people feel safe and they feel they have a fair shot at living a good life, a world where there are mechanisms to help curb our deficiencies and overcome our most base impulses." That's the humanistic ethic talking.

Weisman went on to say, "With so many technologies that are emerging and converging to help us eliminate disease, address climate change, and transform scarcity into abundance across the globe, we have but one thing to do, the very thing we've assembled here to do, create the future." Yet his faith isn't in the tools but in the toolmakers. "It's about us," he says. "It's about people. That's how we'll get there and that's the only way we'll get there. We need to do this work together." Once again, Singularity University promotes exponential technologies for the strict purpose of surmounting humanity's greatest challenges. "The stakes have never been higher," Weisman continued. "You understand that we have disconnected ourselves from nature and failed to understand that we are inextricably connected and depend on each other's wellness to thrive as a whole ... So, everyone must do well if everyone is ultimately going to do great, and that's how we have to look at the world today. We all need to do well for all of us, as a whole, to be doing great." That sounds like the humanistic ethic, my North Star, talking.

Our tools—our technologies—like everything else we do, according to this ethic, ought to be for the purpose of improving human welfare, which always includes the wellbeing of our planet. Recall the tremendous success of the flash XPRIZE response to the Deepwater Horizon explosion in 2010. Its impacts on the ocean, marine life, and shoreline were bad enough, but considering it was the largest marine oil spill in history, things could have been much worse were it not for a bunch of toolmakers who cared about the welfare of our planet. "Disaster is a motivator because empathy is a motivator," Peter Diamandis says, "and empathy is never higher than when the same disaster movie has been playing on TV for over a month."[3] So this wasn't about profit or opportunism. What motivated our toolmaking species to make the tools necessary to deal with this disaster was their humanity.

Many SU graduates are also members of the Abundance Digital Community. It's an online community that allows its members to continually learn and be inspired to do our utmost to make the world a better place for everyone by taking on humanity's greatest challenges. During its 2018 annual 360 Conference, Diamandis interviewed cryptocurrency enthusiast Brock Pierce, who told those entrepreneurs and business leaders in attendance this new means of securing financial transactions represents an "opportunity to go from building a world of *me* ... to create a world of, *we*." He concluded his remarks by explaining the Japanese secret of a happy life, *Ikigai*, "find out what you love, find out what you're good at, find out what the world needs. At the intersection of those three is your life's purpose, and, trust me, the money will follow."

It's easy to be pessimistic these days and hard not to be. We're often let down by the leaders we elect to represent our interests. Meanwhile, the huge problems we face, like global warming, income inequality, the overwhelming costs of housing, education, and healthcare, the rise of global authoritarians, and so on, seem only to be getting worse and we want something or someone to blame.

This is what dualistic thinking facilitates: me good, you bad. This makes us feel good, or, at least, better about ourselves and the world, but at somebody else's expense. The humanistic ethic, on the other hand, requires us to believe in the innate goodness of humanity, in the goodness of others, in the goodness of our toolmaking species and the tools we make, because our tools are an extension of our humanity. Kurzweil says, "No other tool-using animal on Earth has demonstrated the ability to create and retain innovations in their use of tools."[4] So, if we believe in humanity, and in the innate goodness within us all, then we must also believe in the humanity and goodness of our toolmaking behaviors.

    Machines are not the antithesis of humanity. They are a profound expression of it, so long as those wielding them do so in the service of human wellbeing and fulfillment. Far from being heartless, we must realize individuals like Ray Kurzweil, Peter Diamandis, Will Weisman, and Brock Pierce, to paraphrase my Afghanistan friends, "are human beings that have hearts like everyone else," because they are motivated by love and compassion and care, the best the human soul has to offer. This is why I feel so positive about our future, because I believe in toolmakers like these and in the tools they make, and because I know they are motivated to help create a future where "we feed everyone, where we educate everyone, where we shelter everyone, a world where people feel safe and they feel they have a fair shot at living a good life." And this is why

I'm confident when I say, *yes*, I'm still following my North Star.

---

[1] Kurzweil, Ray, *The Age of Spiritual Machines*, Viking Press, New York, NY, 1999. p. 110.
[2] Fromm, Erich, *Man for Himself*, An Owl Book, Henry Holt & Co., New York, NY, 1947, p. 13.
[3] Ibid.
[4] Kurzweil, ibid., p. 23.

AFTERWORD

Although I'm of a generation that tends to prefer *Star Trek* to *Star Wars*, I was nevertheless born on May 4th, which has since become known as *Star Wars* day: "May the *Fourth* be with you." Initially heralded for its unprecedented special effects, the original saga has continued to capture the imaginations of new generations, not merely because of its once groundbreaking visuals, but because of its retelling of an archetypal myth inherent within the human psyche. Stories about the conflict between the Darkness and the Light surely predate recorded history, and are at least as old as the stories of the solar deities previously mentioned: the stories of Attis,

Dionysus, Horus, Mithra, Krishna, and Christ, among others, all said to have returned each year on the winter solstice to defeat the dark and usher in a new era of light.

In *Star Wars*, however, the ultimate goal isn't to defeat the dark, but to bring balance to the galaxy, between the Darkness and the Light, by using the Force. The Force is a mysterious energy that exists everywhere in the Universe and binds everything together. Although it has a will of its own, those who understand it feel at one with the entire Universe, a connection to all within it, and are able to use the Force to do extraordinary things.

This pretty much sums up my understanding of evolution. Evolution is a force that exists everywhere in the Universe: in its endless heavens and upon its endless earths, and within the unfolding minds of their occupants and the things those minds fashion into tools and technology. Evolution also unifies all things through convergence. The many become one. Brahman awakens to discover the myriad of things was only a dream, that the light and the darkness, like all opposites, are part of a whole. Evolution also gives entire species the power to do extraordinary things by enabling them to adapt and be transformed into something new. Today, especially, because the evolution of technology outpaces that of our biology (at least for now), it has also

AFTERWORD

granted each of us the power to do things that would have been considered impossible by our predecessors just a few decades ago. When it comes to advances in areas like AI and machine learning, extended reality, medicine, space exploration, and so forth, we have entered an almost vertical point upon evolution's exponential curve.

That this is so, that evolution occurs, is indisputable, despite those who dismiss it as only a theory. *How* it happens, like most disciplines, must be taught as theory, but this is not a sound reason to argue against evolution. Nobody disputes the existence of music just because musicians learn music theory. That humans must have emerged from earth is ancient wisdom that was understood by our ancestors thousands of years before Charles Darwin published *The Origin of Species* in 1859. The very name, *Adam*, from the familiar Hebrew origin story, means "of the earth." Likewise, the Hopi word *pueblo* refers to both people and the adobe buildings they inhabit because both are made of mud. The word *human* itself means "earth," which is why it shares the same root as *humus*. The ancient Greek story of our origins says Prometheus created humans by mixing earth and water together, similar to stories of people and animals shaped from mud that are found among numerous Native American, Mesoamerican, African, and Chinese legends.

In Babylonian mythology, Apsu is the name of a watery abyss from which all life eventually springs and is the deity who fashions human beings out of mud. In ancient Egyptian mythology life springs from Nun, meaning *Chaos*, "the primordial ocean in which before the creation lay the germs of all things and all beings,"[1] according to *The Larousse Encyclopedia of Mythology*. "They taught that inside Nun, before creation, there had lived a 'spirit, still formless, who bore within him the sum of all existence.'"[2] That sounds a lot like evolutionary theory to me.

Whether we give it names like Prometheus, Apsu, or Nun, evolution has been a force *H. sapiens* has recognized for thousands of years. Even after Darwin consigned this ancestral knowledge to the domain of science, it hasn't been content to remain so tightly confined, no more than *information* itself has been satisfied to remain cramped within our genes. Less than a hundred years after *Origins*, Teilhard de Chardin began forecasting that evolution would eventually culminate in the convergence of human consciousness, which is not unlike Ray Kurzweil's more recent prediction that the entire universe is about to become "saturated with our intelligence."

For some people, these rapid changes seem terrifying, while others are enthusiastic about what they consider is the most transformative period in human existence. Some caution is warranted and will help us advance with fewer regrets, so long as we don't let our biological propensity toward fear prevent us from advancing at all. Not that fear can really halt this force of nature. Like Old Man River, it just keeps on rollin' along; whether we are going with the flow or are terrified by the rapids we encounter, ultimately makes no difference. *The times, they are a changin'*, to cite another song, with or without our permission.

Yet if we are at one with this force, if we are at peace with Evolution's Way, we can foresee where it's taking us and use the adaptive abilities it empowers us with to help shape the future we want. It doesn't have to become the dystopian future many of us dread. Nor is it going to be everything others hope for. Once realized, our Utopian dreams are never the perfect worlds we imagined. They still have many challenges. But they are still better worlds than most ever believed possible.

For the past 13.7 billion years the known Universe has been moving down a path toward exponentially higher states of complexity, consciousness, and unity. This long path has become clear in hindsight and now appears much shorter as we turn toward tomorrow and see it passing right before our astonished eyes. It astonishes us because it is our impending destiny, a destiny based upon the economics of human welfare and growth; a destiny without poverty, hunger, or discrimination; a destiny without the three kinds of suffering even the Buddha learned to live with—old age, disease, and death; a destiny in which we all live in peace, as one community on a healthy planet, even as others are at home on other planets or inside floating space cities; a destiny in which each of us awakens every day to determine what we will do to make the world, nay, the whole Universe a better place. That's the path we are on. That's our trajectory. That's Evolution's Way.

---

[1] Guirand, Felix, ed., *The Larousse Encyclopedia of Mythology*, Aldrington, Richard & Ames, Delano, translators, Barnes & Noble Books, New York, 1959, 1994, p. 11.
[2] Ibid.

ACKNOWLEDGMENTS

It would be an oversight in a book of this nature not to acknowledge all those who have come before me, or all that has come before me. I stand upon the shoulders of giants, some known and some unknown, as well as within the tracks of the smallest, earliest, and most forgotten creatures whose star-born longings remain within the cells and sinews of my own body. My grasp for more understanding is but an evolution of their groping in the dark.

I'm also grateful to the members of Clifton Unitarian Church in Louisville, Kentucky, and of the Unitarian Universalist Church of Spokane, Washington, who have tolerated my musings over the years, including the original sermons this book is largely based on. Their own curiosity has provided me much room to grow and evolve. I shall always remain in their debt.

I'm equally as indebted to all those cited throughout this book, with special appreciation to those who have had a lasting impact upon my life. They are, in alphabetical order, Pierre Teilhard de Chardin, the Catholic priest, mystic, and evolutionary scientist who should be honored and remembered today for the accuracy of his inspirational vision decades before anyone else could have seen what was coming; Peter Diamandis and Steven Kotler, for their lifechanging work, *Abundance: The Future Is Better than You Think,* and to Diamandis, especially, for continuing to inspire and alter my life through Singularity University and the Abundance Digital Community, and for making me a far more optimistic person than I would otherwise have become; theologian and my fellow heretic, Matthew Fox, for first helping me understand the Universe story, making me feel smaller than ever, yet part of something larger than I can comprehend; Erich Fromm, whom, more than anyone, has influenced my thinking and worldview; Ray Kurzweil, for the gift of his genius, the hope he inspires, and future he works toward; Frank L. Tipler, the provocative physicist who helped expand my understanding of what it means to be a person and for teaching me how to love the people of the future; and to

ACKNOWLEGMENTS

Master Morrihei Ueshiba, the founder of Aikido, the Way of Harmony, whose love of nature is the foundation of his teachings about how best to move through life: *irimi* (to face what is before us with courage), *tenkan* (turning to see from other perspectives), *shihonage* (the agility to spin ourselves in new directions), and *suwariwaza* (the strength of being humble and still). Moving through the world in this way has helped me embrace Evolution's Way, what his Shinto religion calls *Kami*, the Spirit of nature that dwells in all things.

BIBLIOGRAPHY

BOOKS & ARTICLES

Angier, Natalie, "All but Ageless, Turtles Face their Biggest Threat: Humans," *New York Times* (Science), December 12, 2006

Armstrong, Karen, *The Battle for God: A History of Fundamentalism*, Random House, New York, NY, 2000, 2001

Asprey, David, *Super Human*, Harper Wave, New York, NY, 2019

Coleman Barks, *Like This*, Library of Congress Catalog # 89-092393, 1990

Berry, Thomas, *The Dream of the Earth*, Sierra Club Books, San Francisco, CA, 1988, 1990

Bricker, Darrell, *Empty Planet*, Crown/Archetype, Kindle Edition.

Brockman, John, ed., *What Are You Optimistic About,* Harper Collins, New York, NY, 2007

Buber, Martin, *I and Thou*, trans. by Walter Kaufmann, Charles Scribner's Sons, U.S., 1970

Bucke, Richard M., *Cosmic Consciousness*, Penguin Books, New York, NY, 1901, 1991

Burton, Robert, *On Being* Certain, St. Martin's Press, New York, NY, 2008

Clark, William R., *Sex and the Origins of Death*, Oxford University Press, New York, NY, 1996

de Chardin, Pierre Teilhard, *The Future of Man*, Harper and Row, New York, NY, 1959, 1964

BIBLIOGRAPHY

Denning, Steve, "Lest We Forget: Why We Had a Financial Crisis," *Forbes* Senior Contributor, Forbes.com, November 22, 2011

Diamandis, Peter H., and, Kotler, Steven, *Abundance: The Future Is Better than You Think*, Free Press, New York, NY, 2012

Diamandis, Peter H., & Kotler, Steven, *Bold: How to Go Big, Create Wealth, and Impact the World*, Simon & Schuster, New York, NY, 2015

Emerson, Ralph Waldo, *The Writings of Ralph Waldo Emerson*, ed. Brooks Atkinson, Random House, Inc., The Modern Library, New York, NY, 1940

Ewalt, David M., *Defying Reality: The Inside Story of the Virtual reality Revolution*, Blue Rider Press, New York, NY, 2018

Flannelly, Kevin J., and Galek, Kathleen, *Religion, Evolution, and Mental Health: Attachment Theory and ETAS Theory,* **Journal of Religion and Health** (2010) 49-337-350, Published online, March 17, 2009, Springer Science+Business Media, LLC, 2009

Fox, Matthew, *One River, Many Wells*, Jeremy P. Tarcher/Putnam, New York, NY, 2000

Fromm, Erich, *Escape from Freedom*, Avon Books, Heart Corporation, New York, NY, 1941, 1965

Fromm, Erich, *Man for Himself*, An Owl Book, Henry Holt & Co., New York, NY, 1947

Fromm, Erich, *The Art of Loving*, Harper & Row, New York, NY, 1956

Fromm, Erich, *The Sane Society,* Henry Holt & Company, New York, NY, 1955

Guirand, Felix, ed., *The Larousse Encyclopedia of Mythology*, Aldrington, Richard & Ames, Delano, translators, Barnes & Noble Books, New York, 1959, 1994

Harari, Yuval Noah, *Sapiens: A Brief History of* Humankind, Harper Collins Publishers, New York, NY, 2015

Harrison, Guy P., *Think Before Your Like*, Prometheus Book, New York, NY, 2017

Harvey, Andrew, *Teachings of Rumi*, Shambhala, Boston & London, 1999

Jain, Naveen, *Moonshots: Creating a World of Abundance*, John August Media, LLC, Kindle Edition

Journal of Scientific Exploration, Vol. 9, No. 2, pp. 223-229, 1995

Kegan, Robert, *The Evolving Self*, Harvard University Press, Cambridge, MA, 1982

Kopp, Sheldon B., *If You Meet the Buddha on the Road, Kill Him!*, Science and Behavior, Palo Alto, CA, 1972, Bantam Books, New York, NY, 1976

Kurzweil, Ray, *The Age of Spiritual Machines*, Viking Press, New York, NY, 1999

Kurzweil, Ray, *The Singularity is Near*, Viking Penguin, New York, NY, 2005

Lehrer, Jonah, *How We Decide*, Mariner Books edition, New York, NY, 2010

Lukianoff, Greg, & Haidt, Jonathan, *The Coddling of the American Mind*, Penguin Press, New York, NY, 2018

MacRae, Allan, & Zehr, Howard, *The Little Book of Family Group Conferences: New Zealand Style*, Good Books, Intercourse, PA, 2004

Marinoff, Lou, *Plato Not Prozac!* HarperCollins, New York, NY, 1999

Mellon, Jim, *Juvenescence: Investing in the age of longevity.* Harriman House. Kindle Edition, 2017.

Mitchell, Stephen, *Tao Te Ching*, Harper & Row Publishers, New York, NY, 1988

BIBLIOGRAPHY

Reese, Byron, *The Fourth Age: Smart Robots, Conscious Computers, and the Future of Humanity*, Atria Books, 2019

Ridley, Matt, *The Rational Optimist*, Harper Collins Publishers, New York, NY, 2010

Rosling, Hans, *Factfulness*, Flatiron Books, New York, NY, 2018

Rumi, *Like This*, versions by Coleman Barks, Library of Congress Catalog # 89-092393, 1990

Russell, Bertrand, *The Art of Philosophizing*, Philosophical Library, New York, NY, 1968

Russell, Peter, *The Global Brain*, J.P. Tarcher, Inc., Distributed by Houghton Mifflin Company, Boston, MA, 1982

Scott, Mark S. M., *Journey Back to God: Origen on the Problem of Evil*, American Academy of Religion (AAR) and Oxford University Press, Inc., New York, NY, 2012

Segal, F. Alan, *Life after Death: The Social Sources*, Chapter 5 of The Resurrection, edited by Stephen Davis, Daniel Kendall, and Gerald O'Collins, Oxford University Press, 1997, 1998

Sinclair, David A., *Lifespan: Why We Age and Why We Don't Have To*, Atria Books, New York, NY, 2019

Singer, Peter, *One World: The Ethics of Globalization*, 2nd ed., Yale University Press, U.S. 2004

Strozier, Charles B., Terman, David, M., Jones, James W., & Boyd, Katherine A., *The Fundamentalist Mindset*, Oxford University Press, New York, NY, 2010

Swimme, Brian, *The Universe is a Green Dragon*, Bear & Company, Inc., Sante Fe, NM, 1984

Tipler, Frank L., *The Physics of Immortality*, An Anchor Book, Doubleday, New York, NY, 1994

Ueshiba, Morrihei, *The Art of Peace*, trans. by John Stevens, Shambhala, Boston, MA, 1992

Underhill, Evelyn, *Mysticism*, Dover Publications, Mineola, NY, 1930, 2002

"Why Americans are putting off doctor visits: Zocdoc," CNBC, Published Tue. June 23, 2015, 2:40 PM

ONLINE SOURCES

www.cbsnews.com/news/the-cost-of-dying-end-of-life-care/

www.commonwealthfund.org/press-release/2015/us-spends-more-health-care-other-high-income-nations-has-lower-life-expectancy

www.ehealthinsurance.com/resources/individual-and-family/much-health-insurance-families-cost

www.elevian.com

https://endhomelessness.org/homelessness-in-america/homelessness-statistics/state-of-homelessness-report/

www.guardian.co.uk/commentisfree/cif-green/2010/jun/18/matt-ridley-rational-optimist-errors

www.kurzweilai.net/the-law-of-accelerating-returns

https://www.minimum-wage.org/articles/history

www.npr.org/templates/story/storyComments.php?storyId=133834740&pageNum=2&pPageNum=2

www.pewresearch.org/fact-tank/2017/12/06/despite-concerns-about-global-democracy-nearly-six-in-ten-countries-are-now-democratic/

www.popsci.com/bacteria-enzyme-plastic-waste

www.rentcafe.com/blog/rental-market/apartment-rent-report/rentcafe-apartment-market-report-june-2018/

www.slate.com/articles/health_and_science/elements/features/2010/blogging_the_periodic_table/aluminum_it_used_to_be_more_precious_than_gold.html

BIBLIOGRAPHY

www.thegatesnotes.com/Books/Development/Africa-Needs-Aid-Not-Flawed-Theories

www.un.org/en/sections/issues-depth/water/

www.wired.com/1995/04/pear/

www.xprize.org

www.youtube.com/watch?v=BUTwZUHJRTI

INDEX

1

10X, 140

3

3-D printer, 97
3-D printers, 76, 120
3-D printing, 120
    3-D printed houses, 220

5

5G, 188

A

Abundance, iv–vi, 147, 156
Abundance 360 Conference, 142
Abundance Digital Community, 147–49, 245
Afghanistan, 241
AI, 250
Alcoa, 154
AliveCor, 222
aluminum, 154–56, 162
American Philosophical Practitioners Association, 135
amygdala, 92, 166, 184
anima, 40
anthropocentric, 14
apocalypticism, 93, 131
Apollo 11, 234
Apollo 7, 234
Apostle Paul, 13, 72
Appleby, Scott, 93
Aristotle, 195
Artificial Intelligence, 115
Asprey, David, 198
augmented reality, 188

B

Barra, Mary, 115
basic quality of life, 219
Berger, Peter, 182
Berry, Thomas, 4, 28
Bezos, Jeff, 119, 147
Big Bang, 20
Bill and Malinda Gates Foundation, 98
biology, 33
BioRegional Development, 160
bipedal animals, 165
Blake, William, 86
Blue Origen, 147
BQL, 219
Brahman, 62, 82, 249
    Brahmic Bliss, 80
Branson, Richard, 147
Brockman, John, 127
Buber, Martin, 41
Bucke, Richard, 79–82
    Cosmic Consciousness, 81
Bucke, Robert, 88
Burton, Robert, 186
Business Ethics, 173

INDEX

C

Cambrian Explosion, 23, 227
Cameron, James, 143
carbon-neutral fuel, 142
Celularity, 203–4
Children, Young Persons and Their Families Act, 146
Clark, William R., 196
CO_2, 120, 145
coal, 120, 161
    Peabody Coal, 120
Cocoon, 205
cogito ergo sum, 189
Cognitive Behavioral Therapy, 135
Compassionate Use Act, 203
complexity, i, 6, 19–26, 86
convergence, 20–26, 62, 111
    of technologies, 25
convergence., 249
cosmology, 16
crowdfunding, 151

D

Darwin, Charles, 228, 250
Davenport, Christian, 147
Deepwater Horizon, 143–44, 244
dematerialization, 118
Denver, John, 136
Depression cakes, 157
Descartes, 189
Diamandis, Peter, iii, 92, 111, 114, 122, 127–28, 139, 143, 146, 151, 160, 174, 244–62
differentiation, 7
dinosaurs, 35
disruption, 114
Dojo, v
driverless cars, 114
drones, 121
dualism, 237
Dunne, Brenda, 46

E

Easter, 65
economics, 210
education, 221
Einstein, Albert, 19, 36, 46
electrolysis, 154
Elevian, 204
Emerson, Ralph Waldo, 236
eschatology, 12
ETAS, 167
eudaimonia, 135
eukaryotic cells, 21, 227
evolution, 20, 84
    convergence, 22
    definition of, i
    evolution of consciousness, iv
    evolution of technology, 86
Evolutionary Psychology, 167
Evolutionary Threat Assessment Systems Theory, 167
Ewalt, David, 181–82
Explorer I, 233
exponential, 112
extended reality, 117–18, 250
Exxon Valdez, 144

F

Family Group Conference, 146
Finette, Pascal, 113
Flannelly, Kevin J., 167
Flint, Michigan, 161
Four Killers, 198
Fox, Matthew, 159
Frankl, Victor, 182
free space, 119
Freedom Dividend, 213
Freud, Sigmund, 83
Fromm, Erich, 83, 133, 211, 242
fundamentalist mindset, 93

G

Gagarin, Yuri, 233

Gates, Gates, 98–100
Gatesnotes.com, 99
GDF11, 204
Gemini program, 234
General Motors, 115
genetically modified organisms, 95–97
Gilbert, Paul, 167
Gildert, Suzanne, 116
Giovannitti, Fred, 144
Goldilocks planet, 20
Google Sky, 86
Google X, 140
Great Depression, 157
Guinness Book of World Records, 150

H

Haidt, Jonathan, 126
Hanson Robotics, 116
Harari, Yuval, 104–6
healthcare industry, 222
Healthy.io, 222
Hepatitis C, 200
Hildegard of Bingen, 80
Hill, Daniel, 94
Hillman, James, 125, 129, 137
hippocampus, 166
holodeck, 189
Homo sapiens, i, 1, 6, 15, 18
Hubble Space Telescope, 234
human population, 90, 176
humane meats, 120
humanistic ethic, 171, 242
Hurricane Maria, 141

I

iGen, 126
Ikigai, 245
illumination, 79
information, 3
Institute for Regenerative Medicine, 77

International Criminal Court, 103
International Monetary Fund, 103
International Space Station, 155, 235
Internet, 10
interstellar space, 235
Irvine, William, 136

J

Jacobstein, Neil, 114
Jain, Naveen, 140, 156, 158, 170
Jane Austen, 113
Jepson, Mary Lou, 190
Jesus, iv, 158
Julius Caesar, 154
Jung, Carl, 7, 183
Jupiter, 234

K

Kaiser Family Foundation, 207
Kasparov, Garry, 53
Kegan, Robert, 60, 182
Kopp, Sheldon B., 130
Kotler, Steven, iii, 92, 127–28, 160
Kurzweil, Ray, iii, 43, 50, 74–76, 86–87, 105, 111, 123, 239, 246, 251
KYOTO Protocol, 103

L

Laika, 233
Lascaux, 181–82, 193
Lehrer, Jonah, 186
Lennon, John, 102
  Imagine, 102
Leonov, Alexei, 233
logic, 134
Lukianoff, Greg, 126, 132
Luna I, 233

INDEX

M

machine learning, 250
MacRae, Allan, 145, 151
Mariner 4, 234
Mariner II, 233
Marinoff, Lou, 135
Mars, 119, 148, 234
Marty, Martin E., 93
Maryniak, Gregg, 119
Masdar, 160
Maslow, Abraham, 212
Massella, Luke, 76
Massively Transformative Purpose, 147
Mechtild of Magdeburg, 80
medical technologies, 222
Meister Eckhart, 6
Mellon, Jim, 199
memes, ii, 4, 230, 232
Mir space station, 234
Moore's law, 220
Morihei Ueshiba, 43
Moses, 185
MTP, 147
multicellular organisms, 22, 35, 227
Musk, Elon, 147, 151, 235

N

Naam, Ramez, 175
NAFTA, 103
Napoleon, 154
NASA, 147, 233
Neptune, 234
Nesse, Randolph, 98
New Zealand Legislature, 146

O

object permanence, 59
Odin, 66
oil, 161
old age, 197

omega point, 6, 15, 23, 62, 82, 84
One Planet Living, 160
Openwater, 190

P

Paris Climate Agreement, 103
Peoc'h, René, 44
PET, 177
Pierce, Brock, 245
Pioneer 10, 234
Pioneer 11, 234
placental stem cells, 203
plastic-eating bacteria, 177
Plato, 191
Pliny the Elder, 154
President Bill Clinton, 122
President Kennedy, 139
Primary Pulmonary Hypertension, 149
Princeton Engineering Anomalies Research, 46
  PEAR, 46
prokaryotes, 20, 34, 226
prolonged vitality, 198

Q

quantum entanglement, 46

R

Ragnarok, 13
rainforests, 2
rationalizations, 133
Raxworthy, Christopher, 196
reason, 133
Reese, Byron, 168–70, 198
Regulation Theory, 94
Restorative Justice, 146, 171
resTORbio, 201–2
Ridley, Matt, 90–92, 95, 99, 127
robot
  meaning of, 39
robots, 115

Rolle, Richard, 80
Rosling, Hans, 127–28
Rothblatt, Martine, 148–51
Rumi, 80, 136
Russell, Bertrand, vii, 189
Russell, Peter, 231, 236

S

salt-based batteries, 141
Samadhi, 80
Samumed, 202
Sanctuary AI, 116
Sanders, Bernie, 213
sanitation, 97
satesphere, 235
Saturn, 234
scarcity, 157, 162
Schmidt Family Foundation, 143
Schmidt, Wendy, 143
Segal, Alan, 68
self driving car, 141
Shepard, Alan, 233
Sinclair, David A, 197
Singer, Peter, 103
Singularity University, iii, 110–12, 123, 241–44
Sir Thomas More, 168, 175
Sirius XM Radio, 148
Skylab, 234
Skysource/Skywater Alliance, 122
sleep apnea, 49
Socrates, 195
solar energy, 121, 175
Sophia, 116
space cities, 119
space exploration, 250
Sputnik, 233
Sputnik II, 233
Stanford University, 222
Star Trek, 142, 189, 248
Star Wars, 248
Starlink, 235
stem cells, 120
Stoicism, 135

Strozier, Charles, 93
super-organisms, 24
Surmet Corporation, 155
sustainability, 159
Swimme, Brian, 28

T

Taoism, 191, 238
technology, ii, 242
Teilhard de Chardin, Pierre, 62, 111, 230, 251
Teller, Astro, 140, 142, 151
Teller, Edward, 140
Terman, David, 94
terrestrial planets, 21
Tesla, 148
The Fundamentalist Project, 93
The Jetsons, 38
The Law of Accelerating Returns, 112
the Singularity, 112
Theobald, Robert, 214–15
theology, 83
therapy, 129
Tiger Toilet, 98
Tipler, Frank, 16, 72
TORC1, 201–2
transcendence, 228
Transcendentalism, 236

U

U.S. homeless rates, 218
Udacity, 221
unicellular organisms, 21, 22
Unitarian Universalism, vi, 61
United Nations, 108, 161
United Therapeutics, 150
Universal Basic Income, 214–16
Universal Declaration of Human Rights, 108
Universe, 5, 12–17, 40, 62, 225–27
Uranus, 234
Utopia, 168

INDEX

V

van Bakel, Rogier, 46
Ventura, Michael, 125, 129, 137
Venus, 234
View-Master, 180
Viking 1, 234
Viking 2, 234
Virgin Galactic, 147
Voyager 1, 234

W

WAMO, 141
waterless toilets, 98
Watson, 52–55, 58–62
   Jennings, Ken, 54
   Jeopardy, 52
Weisman, Will, 243–44
White, Ed, 233

Wilford Brimley, 205
wind energy, 121
wind power, 141
Witherspoon, Jay, 160
WNT pathway, 202
World Bank, 103
World Trade Organization, 103
World War II, 157
World Wildlife Fund, 160

X

XPRIZE, 121, 143–45, 244

Y

Yang, Andrew, 213–16
Yggdrasill, 66